FLIGHT TRAINING ACADEMIC ENHANCER

A useful guide to preventing RETREATING BRAIN STALL

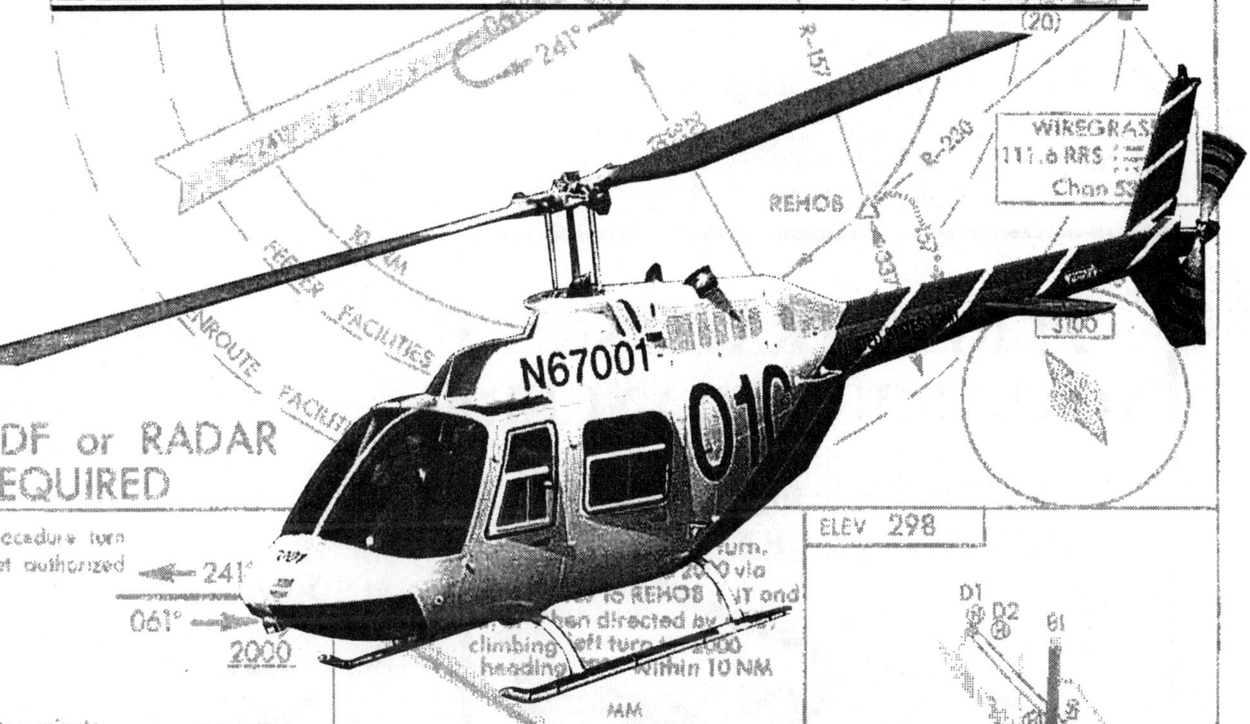

BOOK II - INSTRUMENTS

compiled by: d.w. creech

See us at:
http://www.bandabooks.com

$16.95

Flight Training Academic Enhancer, Book II - Instruments

ISSN 1066-078X
ISBN # 0-89826-073-6

This book is not an official publication of the Department of Defense, the Department of the Army, or the United States Army Aviation Center, nor does its publication in any way imply its endorsement by these agencies.

References previously reviewed for OPSEC considerations.

Printed in the United States of America.

1st Edition
2nd Printing - New Cover Design

Cover photo: TH-67 Creek
Photo Credit: Bell Helicopter

ALSO AVAILABLE ......

Book I of the series

PRIMARY

PREFACE

This study guide was compiled in an effort to enhance the academic potential of students attending the INSTRUMENT portion of the Initial Entry Rotary Wing (IERW) course at Ft Rucker, AL, and as a source of review material for those students that have already completed such training.

There is no substitute for the many hours of studying that will be required to complete this program. It is not the intention of this book to teach you or to replace those hours of study, but to provide an educational tool for achievement-minded IERW students to excel. It is also not the intention of this book to provide an "easy out" or a "crutch" for students that do not have what it takes to become an Aviator in the United States Army. On the contrary, unless you have some grasp of the material presented in academics, many of the questions and answers contained in this book will not make much sense.

This book is not a "cheat sheet." No attempt has been made to emphasize the importance of some questions over others (ie, test questions), and no attempt has been made to duplicate Daily Questions (flight line assignments). In most cases, questions are directly related to the Learning Objectives presented in class.

The vast majority of questions and answers were obtained straight from the Student Handouts presented in academics. These handouts are the source of test questions, so in order to avoid any confusion the use of outside references was kept to a minimum. Additional references, such as the appropriate subject-matter FM or AR, were used only in instances where clarification was necessary.

This book should not be used as an alternative to the appropriate Student Handouts and notes taken during class. Instead use it as a study supplement. Keep in mind that information is constantly being revised and incorporated into an increasingly sophisticated and rapidly changing training program. I will make every attempt to keep this book up-to-date but the responsibility will always fall on you to keep up with the changes that affect you and your abilities as an aviator.

ACKNOWLEDGEMENTS

I accept credit only for the time-consuming compilation of the material contained herein. The majority of the credit goes to the fine academic instructors, instructor pilots, and support personnel that make this training program the best in the world.

STUDY GUIDE
TABLE OF CONTENTS

STUDY GUIDE
TABLE OF CONTENTS

TOPIC	LOCATION

ATTITUDE INSTRUMENT FLYING/BASIC INSTRUMENTS

Q. What is meant by the "longitudinal axis" of an aircraft?
A. It is an imaginary line drawn down the center of the aircraft from nose to tail and on which the aircraft may freely rotate. When an aircraft rotates about its longitudinal axis it is said to be banking (or rolling).

Q. What is meant by the "lateral axis" of an aircraft?
A. It is an imaginary line drawn from wingtip to wingtip (side-to-side in a helicopter) and on which the aircraft may freely rotate. When an aircraft rotates about its lateral axis it is said to be pitching (nose-up or nose-down).

Q. What is meant by the "vertical axis" of a aircraft?
A. It is an imaginary line generally entering the bottom of the aircraft at the center of gravity (CG), exiting the top of the aircraft along the same line and on which the aircraft may freely rotate. When an aircraft rotates about its vertical axis it is said to be yawing (nose left or right; slipping or skidding).

NOTE: Aviators often refer to cyclic movements in terms of "fore, aft, and lateral." In this instance a "lateral" movement is a displacement of the cyclic in the lateral (left or right) direction and should not be confused with rotation about the lateral axis. In actuality, "fore and aft" movements are about the lateral axis while a "lateral" movement is about the longitudinal axis.

Q. The attitude of an aircraft in flight is described in terms of pitch, roll (bank) and yaw. Describe each of these terms.
A. Pitch - the angular relationship of the longitudinal axis to the actual horizon. An aircraft's pitch attitude revolves around the lateral axis.
Roll - the angular relationship of the lateral axis to the actual horizon. An aircraft's roll attitude revolves around the longitudinal axis.
Yaw - the angular relationship of the longitudinal or lateral axes to the direction of flight. An aircraft's yaw attitude revolves around the vertical axis.

Q. What is the <u>primary</u> instrument for determining pitch attitude?
A. The attitude indicator (commonly called the "artificial horizon").

Q. The attitude indicator is scaled to indicate degrees of pitch above or below the "horizon." Describe this scale.
A. A vertical line in the center of the instrument is incrementally marked with horizontal lines perpendicular to the vertical line and separated by tick marks. Each horizontal line represents 10° of pitch while the smaller tick marks represent 5° of pitch.

Q. In addition to the attitude indicator, what other instruments may be used for interpreting pitch attitude?
A. Altimeter, Vertical Speed Indicator (VSI), and Airspeed indicator.

Q. Assuming that power is held constant, describe the relationship between pitch attitude change and the three above-mentioned instruments.
A. <u>Pitch up</u> <u>Pitch down</u>

Altimeter	increase	decrease
VSI	indicates climb	indicates descent
A/S indicator	decrease	increase

Q. What is the difference between a VSI and an IVSI?
A. A VSI has 6 to 9 seconds of lag, whereas the Instantaneous Vertical Speed Indicator (IVSI) has no apparent lag.

Q. What is the <u>primary</u> instrument for determining bank attitude?
A. The attitude indicator.

Q. In addition to being scaled to indicate degrees of pitch above or below the horizon, the attitude indicator is also scaled to indicate degrees of bank left or right. Describe this scale.
A. Each "tick" mark on either side of vertical indicates 10° of roll left or right of the horizon.

Q. In addition to the attitude indicator, what other instruments may be used for interpreting bank attitude?
A. Heading indicator or RMI, turn-and-slip indicator, and magnetic compass.

Q. Describe the relationship between bank attitude change and the three above-mentioned instruments.
A. The Radio Magnetic Indicator (RMI) will show heading changes for turns left or right; the turn-and-slip indicator has a needle that indicates both the direction and rate of turn; and the magnetic compass, like the RMI, will reflect heading changes for turns left or right.

Q. The yaw attitude of an aircraft is an important factor in maintaining coordinated flight. What instrument is used for determining whether an aircraft is in coordinated flight?
A. The turn-and-slip indicator.

Q. What is a "slip" and how is it indicated on the instruments?
A. Slipping is when the rate of turn is too slow for the angle of bank; on the turn-and-slip indicator, the ball will be off-centered and to the <u>inside</u> of the turn.

Q. What is a "skid" and how is it indicated on the instruments?
A. Skidding is when the rate of turn is too fast for the angle of bank; the ball will be off-centered and to the <u>outside</u> of the turn.

Q. In coordinated or balanced flight the ball will be centered in the turn-and-slip indicator. This is known as flying in ______.
A. Trim.

Q. What are some factors that affect rate of turn?
A. Angle of bank and True Airspeed (TAS). In addition, the required turning radius or the number of degrees to be turned will determine the necessary rate

of turn.

Q. Rates of turn can be classified into one of three categories. What are they?
A. Standard rate of turn, half standard rate turn, and steep turn.

Q. What is the rule of thumb for determining the bank angle required to obtain a standard rate turn?
A. Generally speaking, about 15% of TAS.

Q. For true airspeeds of 70 to 90 knots, what would be the angle of bank required to establish a standard rate turn?
A. Approximately 12° to 15°, as read from the attitude indicator.

Q. A standard rate turn will result in a rate of turn of ____° per second.
A. 3

NOTE: A standard rate turn is also known as a "2-minute" turn since this rate of turn will result in a complete circle (360°) in two minutes or a course reversal (180°) in one minute.

Q. When should a standard rate of turn be used during instrument flight?
A. For heading changes of 20° or more.

Q. For true airspeeds of 70 to 90 knots, what would be the angle of bank required to establish a half standard rate turn? For a TAS of 120 knots?
A. Approximately 6° to 7½°; 120 x .15 = 18°/2 (for half-standard) = 9°.

Q. A half standard rate turn will result in a rate of turn of ____° per second.
A. 1½

NOTE: A half standard rate turn is also known as a "4-minute" turn since this rate of turn will result in a complete circle (360°) in four minutes or a course reversal (180°) in two minutes.

Q. When should a half standard rate turn be used during instrument flight?
A. For heading changes of less than 20° (also when advised by ATC for Ground Controlled Approaches, etc).

Q. What are the characteristics of a steep turn?
A. Anything that exceeds the characteristics of a standard rate turn.

Q. What are the two primary elements of control in all aircraft?
A. Power and attitude.

Q. What gauge in the helicopter tells us what our "power" level is?
A. Torque (TQ) gauge.

Q. What feature on the helicopter allows us to control power (TQ)?
A. The collective.

Q. With no outside input to compensate, how will power changes affect pitch attitude?
A. Add power (pull up on collective) - nose will pitch up.
Reduce power (push down on collective) - nose will pitch down.

Q. How do you compensate for power adjustments in the helicopter?
A. Adjust cyclic as required to control pitch; adjust pedals as required to adjust yaw.

Q. What is the general rule of thumb for power (TQ) requirements during climbs and descents?
A. 1 psi of torque = 100 fpm.

Q. If power (TQ) levels remain the same, what effect will attitude changes have on the aircraft?
A. Nose down - increase in airspeed (acceleration); decrease in altitude.
Nose up - decrease in airspeed (deceleration); increase in altitude.
Banks left/right - increase in airspeed due to ensuing nose down attitude; decrease in altitude.

NOTE: As the angle of bank of an aircraft changes, the lift vector shifts horizontally requiring power and attitude adjustments to maintain altitude and airspeed.

Q. If attitude remains the same, what will be the effect of power changes on the aircraft?
A. Add power (TQ) - acceleration.
Reduce power (TQ) - deceleration.

NOTE: Keep in mind that in order to maintain attitude as power is increased or decreased an appropriate input will have to be made with the cyclic.

Q. What is the general rule of thumb relating power changes to airspeed changes?
A. 1 psi of torque = 5 kias.

Q. With no outside input to compensate, how will power changes affect yaw attitude?
A. Add power (TQ) - aircraft yaws to the right (nose right).
Reduce power (TQ) - aircraft yaws to the left (nose left).

Q. How do we normally compensate for the above power changes?
A. Add power (TQ) - nose right => apply left pedal.
Reduce power (TQ) - nose left => apply right pedal.

Q. What are the two primary steps used to perform attitude instrument flight?
A. Cross-check and instrument interpretation.

Q. What is meant by cross-check?
A. Scanning and monitoring two or more instruments to determine the attitude and

performance of the aircraft.

Q. What are the three most common errors in performing an instrument cross-check?
A. (F)ixation - staring at one instrument too long.
(O)mission - inadvertently omitting an instrument from the cross-check.
(E)mphasis - concentrating too heavily on one or two instruments during the cross-check.

The following information, though not covered in the INSTRUMENT portion of flight training, is an important part of instrument flying and is presented here as a review.

MAGNETIC COMPASS

Q. What is the power source for the magnetic compass?
A. The earth's magnetic field.

Q. What are the two major components of the mag compass?
A. The case and the float assembly.

Q. What is the vertical white line on the glass face called?
A. A lubber line.

Q. The case of the mag compass is filled with a fluid. What is this fluid and what is its purpose?
A. Varnoline (an acid-free white kerosene); provides buoyancy, lubricates and prevents rust, and dampens oscillations.

Q. The upper portion of the mag compass contains small magnets with adjusting screws to minimize deviation. What is this portion of the mag compass called?
A. The compensator assembly.

Q. A compass card is attached to the float assembly. Letters on the float assembly indicate the cardinal points: N, S, E, and W. Markings are placed every ______° and numbered every ______°.
A. 5; 30.

Q. What is the difference between "variation" and "deviation"?
A. Variation refers to the difference between magnetic heading and true heading, while deviation refers to the difference between magnetic heading and compass heading.

Q. What causes variation?
A. The differences in location between magnetic north and true north.

Q. What causes deviation?
A. Magnetic attractions near the compass (ie, pitot heater, generators).

Q. If you were standing anywhere along an _________ line and you could "see" the north pole and magnetic north, the poles would be perfectly aligned and you would have 0° variation.
A. Agonic.

Q. Magnetic variations east and west of an agonic line are known as what?
A. Isogonic lines (connect points of equal magnetic variation).

Q. What memory aid can be used for remembering how to use isogonic lines to compensate for variation?
A. "East is least" -- subtract variation
"West is best" -- add variation

Example: An aviator flying with a true heading of 080° with a variation in his/her area of 7° E would fly a magnetic heading of 073° (080° - 007°).

Q. Deviation is minimized through mechanical compensation (adjustment screws -- maintenance personnel only). Any deviation(s) in the mag compass that cannot be compensated must be noted on a card which is kept near the compass. What is this card called? How often must this card be updated?
A. Compass correction or compass deviation card; every 12 months.

Q. Acceleration and deceleration can cause erroneous indications on the mag compass. These are most apparent when flying on what headings? What causes this?
A. East and West; magnetic dip.

Q. Acceleration causes the compass to indicate a turn to the ________, while deceleration causes the compass to indicate a turn to the ________.
A. North; south.

Q. What acronym can be used to remember the above condition?
A. A-N-D-S (Accelerate-North; Decelerate-South).

Q. A turning error may be induced in the compass when the aircraft is in a bank on or near what headings? What causes this?
A. North or south; magnetic dip.

Q. Turn error will always be equal to what?
A. Latitude.

Q. At what two points in a 360° turn will turn error be zero?
A. East and west.

Q. Where on the earth's surface will there be almost a 90° turn error?
A. In polar areas where the dip angle exceeds 84°.

Q. What angle of bank should be used when relying on the magnetic compass for

turns to compass headings?
A. 12° to 18°.

Q. When turning to a heading of north, when should you begin your rollout?
A. Prior to reaching the desired heading -- by number of degrees equal to latitude plus half the angle of bank.

Q. When turning to a heading of south, when should you begin your rollout?
A. After reaching the desired heading -- by number of degrees equal to latitude minus half of bank angle.

Q. When turning to a heading of east or west, when should you begin your rollout?
A. Half the angle of bank before reaching the desired heading.

Q. What will the mag compass do when turning from a heading of north?
A. It will briefly indicate a turn in the opposite direction, then in the proper direction but lags behind the turn.

Q. What will the mag compass do when turning from a heading of south?
A. It will indicate a turn in the proper direction but at a more rapid rate than is actually being made (leads).

NOTE: During instrument training you will be required to demonstrate proficiency using the magnetic compass by executing turns with a simulated "stuck" card (fixed RMI). In order to successfully accomplish this maneuver it is important to understand the effects of magnetic dip on the magnetic compass.

As an example, you are flying a heading of 270° and are asked to execute a right turn to due north using only the magnetic compass for heading information.

Since you know that for turns to a heading of north you must roll out prior to reaching the desired heading by an amount equal to the latitude plus half the degree of bank angle (for flights in the vicinity of Ft. Rucker this would be: 31° + (15/2) or approximately 38-39°). To execute this maneuver you would begin a standard rate turn to the right and maintain the rate of turn until the magnetic compass indicated 321-322° (38-39° prior to desired heading). Roll out and the magnetic compass should settle down to approximately due north.

Before executing such a maneuver you'll have to conduct a little mental arithmetic and it is a good idea to have some of the numbers already memorized. Here are the numbers for the Ft. Rucker area:

Turns to north	38-39°	prior to desired heading
Turns to south	23-24°	after desired heading
Turns to east or west ...	7-8°	prior to desired heading

(Good round numbers are 40°, 25°, and 8°)

If you experience problems with the recommended procedure above try the following technique as a "backup" (remember that this is only a technique):

NOTE: This technique is based on the fact that a standard rate turn will result in a complete 360° turn in two minutes; thus, turns to a heading may be timed so long as a standard rate of turn is maintained.

In a standard rate turn the aircraft will turn 45° every 15 seconds. Given this, use the RMI to "picture" 15 seconds at the 45° tick-mark, 30 seconds at the 90° tick-mark, and so on as illustrated in Figure 1.

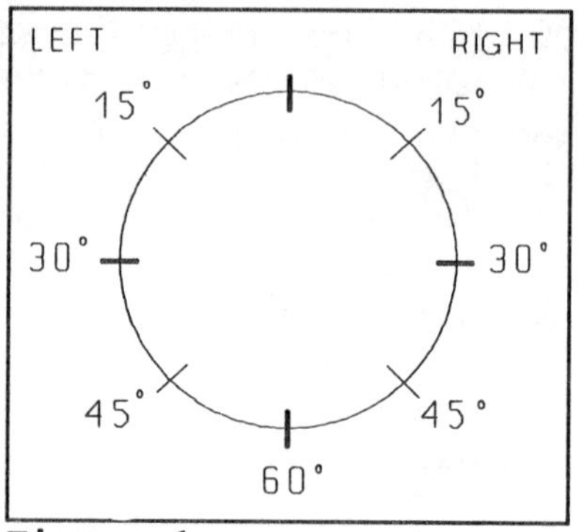

Figure 1

Using Figure 2 and the example above you can quickly see that in order to turn from the current heading of 270° to a desired heading of 360° you will need to execute a standard rate turn to the right for 30 seconds.

A turn to a heading of 135° could be accomplished by executing a standard rate turn to the left for 45 seconds. Some estimation will have to be made for headings between the tick-marks, keeping in mind that every 10° equals approximately 3 seconds.

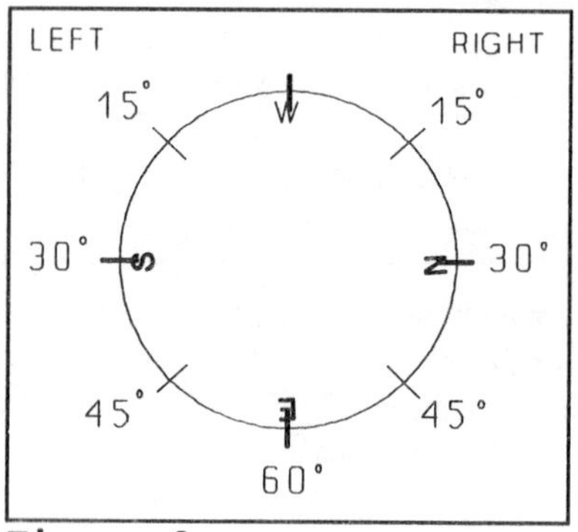

Figure 2

The important thing to remember using this technique is to maintain a standard rate turn throughout the maneuver, note the time as you enter the turn, and roll out at the appropriate time.

If the RMI is covered for this maneuver, use the wind-face side of your E-6B. Place your current heading (as indicated by the magnetic compass) straight up and "visualize" the time to your desired heading.

WIND EFFECT AND VARIATION-DEVIATION

Q. Define the following terms:
A. True Course (TC) -- the proposed flight path measured clockwise from true north.
Drift Correction (DC) -- the correction applied to prevent drifting off-course.
True Heading (TH) -- the angle measured clockwise from true north to the nose of the aircraft.
Track (TR) -- the actual flight path over the ground.
Drift angle (DA) -- the difference between TH and TR.

Q. If enough drift correction is applied your Track will be equal to what?
A. True Course.

Q. Aircraft speed as read from the airspeed indicator is known as what? Is this

value affected by wind?
A. Indicated Airspeed (IAS); no.

Q. What is True Airspeed (TAS)? Is this value affected by wind?
A. Calibrated airspeed corrected for altitude/temperature error; no.

Q. What is Ground Speed (GS)? Is this value affected by wind?
A. The speed of the aircraft over the ground; yes.

Q. Which of the above (IAS, TAS, GS) is used to calculate the time required to fly a given distance?
A. GS.

Q. What is the formula for Ground Speed?
A. TAS +/- wind speed.

Q. Winds from what direction are considered "headwinds"? "Tailwinds"?
A. Headwinds: winds +/- 90° from the aircraft nose (reduces ground speed).
Tailwinds: winds +/- 90° from the aircraft tail (increases ground speed).

Q. How many agonic lines are there in the United States?
A. One.

Q. What is the rule for applying variation when determining magnetic heading (MH)?
A. Take the mid-point of the flight leg and use the nearest whole degree isogonic value. Subtract variations having an easterly value and add variations having a westerly value.

Q. The compass deviation card is divided into two columns. What are they?
A. TO FLY and STEER.

Q. Each column and heading (see above question) is spaced how many degrees apart?
A. 15°.

Q. The value obtained from the compass deviation card is used to compute _______ from the magnetic heading.
A. Compass Heading (CH).

Q. What is the complete formula for determining Compass Heading (CH)?
A. TC (+/-) DC = TH (+/-) VAR = MH (+/-) DEV = CH

NAVIGATION COMPUTER SLIDE RULE

Q. The CPU-26A/P Air Navigation Computer is more commonly known as what?
A. An E6B or "whiz wheel".

Q. The circular slide rule portion of the E6B has a stationary outer and a rotating inner scale. What are these scales called?

A. Outer scale -- miles, distance, or quantity scale.
Inner scale -- time or rate scale.

Q. What distance conversion indexes are available on the outer scale?
A. Nautical miles (NAUT), statute miles (STAT), and kilometers (KM).

Q. The hour index is located at 60 on the inner scale and is represented by what?
A. A large black arrow.

Q. The seconds index is located at 36 on the inner scale and is represented by what?
A. A small black arrow.

Q. Why is the seconds index located at 36 on the inner scale?
A. This divides one hour into 3,600 seconds.

Q. When would you use the 36 index?
A. When time must be calculated in seconds and minutes instead of minutes and hours.

Q. What rule-of-thumb should you remember to help determine whether or not to use the 36 index?
A. If the distance is less than 1/10 of your airspeed, use the 36 index.

Q. For what computations are the inner scale windows used?
A. True altitude, true airspeed, and density altitude.

Q. When using this side of the navigation computer how many factors must be known in order to solve for the unknown?
A. Three (except for density altitude computations).

Q. For density altitude computations accurate results can only be obtained by using pressure altitude (PA). What can you do while in flight to determine your pressure altitude?
A. Set value in Kollsman window to 29.92 and read pressure altitude directly from altimeter.

Q. The computer is labeled CAL ALT on the inner scale. Will the use of indicated altitude in computations rather than calibrated altitude cause any problems?
A. No, this will cause very small and negligible differences in true altitude for most temperature conditions and altitudes that Army aircraft fly.

Q. What is the "rule of 60" in regards to off-course correction?
A. The rule of 60 states that an aircraft headed 1 degree off course will be approximately 1 mile off course for each 60 miles flown.

Q. It follows then, from the rule of 60, that for each mile off course ______ degree(s) of correction must be applied to parallel the course.
A. one.

Q. What are the two rules to remember when making course correction computations?
A. (1) If the distance flown is less than 60 NM, the correction to parallel the TC will be greater in degrees than the number of miles off course. If the distance to fly is less than 60 NM, the correction to converge will be greater in degrees than the number of miles off course.

(2) If the distance flown is greater than 60 NM, the correction to parallel the TC will be less in degrees than the number of miles off course. If the distance to fly is greater than 60 NM, the correction to converge will be less in degrees than the number of miles off course.

NAVIGATION COMPUTER WIND FACE

Q. Air navigation computer slide rule computations require 3 known factors to determine a 4th unknown factor. How does this differ from wind face computations?
A. 4 known factors are required to determine the 5th and/or 6th unknown factors.

Q. The "grommet", located in the center of the transparent plotting disc, is always used to locate what value?
A. Ground speed (GS).

Q. In wind face computations True Airspeed will always be found where?
A. Under the "dot" (made with pencil).

Q. Concentric arcs on the sliding grid are spaced two units apart and are used for ________ calculations.
A. Speed.

Q. Radials left and right of the center heading line are used for calculating what?
A. Drift correction (DC) right and left of true index.

Q. What six computations do we use the wind face portion of the computer for (assuming four of the six are known)?
A. WD, WS, TC/TR, GS, TH, and TAS.

Q. Wind direction (WD) and wind speed (WS) make up a value known as what?
A. Wind velocity (W/V) ==> velocity is a vector quantity having magnitude (speed) and direction.

ENROUTE LOW ALTITUDE (ELA) CHARTS
and
STANDARD INSTRUMENT DEPARTURES (SIDs)

Q. The front cover of an ELA chart consists of a top and bottom section. Among the items of information included here are chart and panel numbers, effective dates, altitude coverage, a pictorial diagram showing area of coverage, and information regarding area charts. How are charts numbered and what is the purpose of having panel numbers?
A. Charts are numbered 1 through 28 for the 48 contiguous states and preceded by the letter "L" (denoting Low altitude charts). Each "chart", in effect, contains two charts - one on each side of the publication. Charts 1 and 2 will be together, as will 3 and 4, and so on. The chart numbers will be indicated on the top section of the front cover and on the inner cover (legend) with an arrow corresponding to the appropriate side for that chart number. The bottom section of the front cover contains an outline of the United States (or area/areas covered by the chart) with an overlay of boxes. These "boxes" are numbered with the ELA chart corresponding to that area. Under the chart numbers are panel numbers (actually letters). Odd numbered charts will have panels "A-B-C-D" and even numbered charts will have panels "E-F-G-H". Panel letters come in handy when attempting to locate an airfield by referencing the IFR Supplement. The last line in the heading for any airfield will give its location on IFR charts (ie, CAIRNS AAF,........., H-4H-5D,L-18F-19A). This example quickly tells us that we can find Cairns AAF on Enroute High Altitude charts H-4 (panel H), H-5 (panel D), and Enroute Low Altitude charts L-18 (panel F) and L-19 (panel A).

Q. What additional information will you find in the top and bottom corners of each panel to help in locating the area of coverage?
A. A city's name.

Q. The first question in this section lists several items of information contained on the front cover of ELA charts. What additional item of information will always be found on the top section of the front cover?
A. Military Training Routes, printed in brown, and listing the route numbers and corresponding altitude ranges.

Q. Can you use an ELA for planned flights in Class A airspace?
A. No. Class A airspace begins at 18,000' MSL. ELA charts are for use up to but not including 18,000' MSL.

Q. How often are ELA charts published?
A. Every eight weeks.

Q. What does FLIP stand for?
A. **FL**ight **I**nformation **P**ublication.

Q. On ELA charts, 1 inch is equivalent to how many nautical miles?
A. This value varies. Use the scales provided at the top and/or bottom of the chart. Also look under the chart number on the front cover. Here you will see something like 1" = 12 NM. Be aware that it is not always the same.

ENROUTE LOW ALTITUDE (ELA) CHARTS
and
STANDARD INSTRUMENT DEPARTURES (SIDs)

Q. As mentioned earlier, the bottom section of the front cover of the ELA has a pictorial representation of the area/areas covered. Cities on this diagram are either in green or black. What is the difference?
A. Cities appearing in green are to assist the user in ORIENTATION and selection of the appropriate charts. Cities in black indicate major hubs for which AREA CHARTS are provided.

Q. What is the purpose of an AREA CHART?
A. These charts portray the airway system and related data required for IFR operations in selected terminal areas at altitudes below 18,000' MSL. Twelve variable scale charts are printed on one sheet.

Q. How often are Area Charts printed?
A. Every eight weeks.

Q. How are time zones indicated on the ELA chart?
A. By a dotted green line (on the front cover pictorial and within the charts themselves).

Q. What agency is responsible for the format and content of an IFR chart?
A. The Federal Aviation Administration (FAA).

Q. What agency is responsible for the accuracy of an IFR chart?
A. The National Oceanic and Atmospheric Administration (NOAA, pronounced like "NOAH").

Q. What horizontal datum is used for ELA charts (United States)?
A. The North American Datum of 1983.

Q. Airfields on the ELA chart are color coded and will be portrayed as either blue, green, or brown. Explain the difference between these.
A. Airfields shown in blue or green have approved low altitude instrument approach procedures published. Those in blue have approved approach procedures published in DOD FLIPs. Airfields shown in brown do not have published instrument approach procedures.

Q. Where should you look to find an approved low altitude instrument approach procedure for an airfield shown in green?
A. In a U.S. Gov't Terminal Procedures FLIP.

NOTE: The U.S. Gov't Terminal Procedures FLIP has an open binding with pre-punched holes in the top. The pages are loose-leaf but generally come with a flexible metal binding strap to keep the pages together. The publication contains instrument approach procedures, airport diagrams, SIDs, and STARs. There are two black bands near the top of each volume and should not be confused with the DOD Terminal FLIPs which are published in two volumes (East/West), have a closed binding at the top and a black band at the top and bottom of each volume.

Q. Where would you look to obtain airport information for an airfield shown in green?
A. In a U.S. Gov't Airport/Facilities Directory (A/FD) (similar to IFR Supplement except that they are published in 7 volumes for the conterminous U.S. and are green with a black band along the binding).

Q. How do you tell the difference between a military airfield and a civilian airfield on the ELA chart?
A. A military airfield is portrayed as two concentric circles (circle within a circle). A civilian airfield is portrayed by a single circle with "cogs" or "ticks". A civil-military airfield is portrayed by concentric circles with "ticks".

NOTE: "Cogs" or "ticks" around the basic airport symbol on VFR Sectionals indicate that services are available during normal working hours. This is not the case for airport symbols on the ELA. Check the IFR Supplement for availability of services.

Q. How is a heliport designated?
A. By an "H" enclosed in a circle.

Q. Somewhere close to each airfield, printed either in blue, green, or brown, will be the airfield name along with related information. If the airport name has parentheses around it, what does this mean?
A. No military landing rights available (prior permission must be obtained from the airfield management before landing or the government may be liable or airfield landing/parking fees).

Q. What is indicated by a solid line box enclosing the airfield name?
A. This indicates that FAR 93 Special Requirements exist. Check Directory/Supplement.

Q. What does a [C] or [D] following the airfield name indicate?
A. Class C or Class D airspace, respectively.

Q. What does a star following an airfield name indicate?
A. Used to indicate part-time operation of a Class C or Class D airfield.

Q. Where can you look to find the hours of operation for Class C or Class D airfields that operate noncontinuously?
A. In the IFR Supplement or inside the ELA chart under Air-Ground (A/G) Tabulation.

Q. On the A/G Communication Tabulation inside the ELA chart, what would the "#" symbol adjacent to a tower name indicate?

ENROUTE LOW ALTITUDE (ELA) CHARTS and STANDARD INSTRUMENT DEPARTURES (SIDs)

A. The tower is closed until further notice.

Q. Is airport elevation given in feet AGL or MSL?
A. MSL (don't be tricked by this question - how can an airfield be above ground level?).

Q. Following the airfield elevation will be runway lighting information. What symbol will occupy this space for the following capabilities: no lighting available, lighting available, and pilot controlled lighting?
A. No lighting available will be indicated by a "-"; lighting available will be indicated by an "L"; and pilot controlled lighting is indicated by an "L" enclosed in a circle.

Q. What does an asterisk following the lighting symbol indicate?
A. Lighting is part-time or on request.

Q. How does a pilot control "pilot-controlled" lighting?
A. By tuning up the airfield frequency and keying the mike 3, 5 or 7 times in five seconds (depending on desired illumination). Lights will normally remain illuminated for 15 minutes from the most recent time of activation.

Q. You see the following line under an airport name: "194 L 46s". What is the length of the shortest runway?
A. For airfields with multiple runways the length of the shortest runway cannot be determined by looking solely at the ELA chart. The value "46" represents the length of the airfield's <u>longest</u> runway in hundreds of feet (to the nearest 100 feet) - in this case, 4600'. The "s" after 46 simply indicates a soft surface runway. To find the length of the shortest runway look in the IFR or VFR Supplement.

Q. An airfield having a runway length of 3465' would be reported as _____ on the ELA chart.
A. 34. 70' is the cutoff for rounding to the nearest 100 (ie, a runway length of 3470' would be reported as 35).

Q. On the ELA chart, how does an airfield indicate that no Special VFR flight is authorized?
A. By the words "NO SVFR" printed above the airfield name.

Q. What does the following line indicate under an airfield name: "(A) * 118.95 327.5"?
A. Automatic Terminal Information Service (ATIS) is available on a part-time basis (*) on VHF 118.95 or UHF 327.5.

Q. For an ATIS that is listed as part-time, where would you look to determine the hours of operation?
A. In the IFR Supplement.

ENROUTE LOW ALTITUDE (ELA) CHARTS
and
STANDARD INSTRUMENT DEPARTURES (SIDs)

Q. What does the following line indicate under an airfield name: "(A) 121.15 ARR 132.1 DEP"?
A. ATIS is available on a continuous basis. Refer to 121.15 for ATIS arrival information and 132.1 for ATIS departure information. Most often used at heavily congested airfields to further reduce controller workload by providing separate arrival- and departure-specific information.

Q. On the ELA chart, all VORs will be depicted in _________ (color). NDBs will be depicted in ___________ (color).
A. Black; black or brown. NDBs navigable by Army helicopters will be depicted in brown and represent low frequency (L/F) or medium frequency (M/F) radio beacons. NDBs depicted in black are UHF beacons and out of the frequency range of our ADF receivers (used by other branches of service).

Q. Navaids will have an associated communication box that gives navaid-specific information such as navaid name, frequency, identifier, coordinates, etc. These communication boxes will be depicted in one of two colors. What are they?
A. Black for VORs (or UHF NDBs) and brown for NDBs (L/F or M/F).

Q. What is meant by a navaid frequency that is underlined in the communication box?
A. No voice transmitted on that frequency. When you tune in to most navaids all you will hear are the familiar dots and dashes necessary to properly identify the station. However, since the navaid is in effect a radio transmitter, voice is often transmitted as well. Some navaids transmit weather information and others are used by flight service for remoted communications outlets (RCO). If you are in a "lost commo" situation it could be that just your radio <u>receivers</u> are malfunctioning. In this instance you would be able to transmit on normal frequencies and "listen in" on a nearby navaid using the navaid receiver, normally a VOR (of course, your receiving station must be informed that you are listening on such and such a VOR so that they will know to transmit on that frequency). <u>Be aware that if the frequency is underlined then this is not an option.</u>

Q. A VOR communication box gives the following information:

```
SIMMONS (T)
109.8   FBG
```

What is the VOR name?
What is it's identifier?
What frequency should the VOR receiver be tuned to in order for navigation equipment to proper utilize the navaid?
What does the symbol (T) indicate?
A. SIMMONS; FBG (of course, the appropriate dots and dashes would also be depicted for proper identification); 109.8 (to identify the station or to listen to voice transmissions over the navaid ensure that the appropriate ICS switch is

in the UP position); the symbol (T) indicates that the VOR is a Terminal Class VOR with a range of approximately 25 NM (see Chpt 5 of this study guide or the IFR Supplement for VOR classifications).

NOTE: If no symbol follows the name it has a range of at least 40 NM. See IFR Supplement for specific classification and range.

Q. What does it mean if any of the information in the communication box is overprinted (has light hash marks through it)?
A. This indicates that the affected data has an abnormal status - check NOTAMS.

Q. What does the symbol "■" in the upper left corner of a communication box indicate?
A. This symbol indicates that Hazardous Inflight Weather Advisory Service (HIWAS) information is transmitted over that VOR when weather conditions warrant. If you are within range of such a navaid and suspect adverse weather it would be a good idea to listen in on the navaid frequency.

NOTE: The "HIWAS" symbol is also used in communication boxes on VFR sectionals, but is located in the lower right corner instead of the upper left corner.

Q. A communication box for a VOR in or near your route of flight has a small black circle with a white "T" inside (negative symbology) and located in the upper right corner of the box. What does this mean and how can it help you?
A. This symbol indicates that a Transcribed Weather Broadcast (TWEB) is transmitted over that VOR. Similar to ATIS information, a TWEB is a pre-recorded, repeating message containing weather information for that area. This can be a valuable tool for the aviator since weather conditions are subject to change and may or may not be the same as forecast weather received prior to departure. Also important, particularly for IFR flight, are updated altimeter settings which can change rapidly with unstable weather conditions.

NOTE: This same symbol is used on VFR Sectionals and is similarly located on communications boxes.

Q. What is indicated by a star preceding a navaid frequency?
A. This indicates that the navaid is operational only part time or on request.

Q. Some communication boxes are depicted so that they appear to have a shadow behind them. Oddly enough, these communication boxes are called SHADOW boxes. When you see such a shadow box associated with a navaid what should immediately come to mind about that navaid?
A. There is a Flight Service Station associated with that navaid. This is important to know because you may have to open or close a flight plan with flight service. Flight Service can also assist a disoriented pilot, provide weather briefings and NOTAMS, and in some cases, Local Airport Advisories (LAA). There are many ways that a Flight Service Station can be of use to an aviator. Use all

available resources!

Q. On what frequencies can I normally contact a Flight Service Station?
A. In addition to the emergency frequencies, 121.5 and 243.0, all Flight Service Stations normally monitor the frequencies 122.2 and 255.4. Other frequencies may be available - if so, they will be shown above the shadow box.

Q. A VOR has the following shadow box information (shown here without the shadow):

```
      122.4    122.1R
 ┌──────────────────────┐
 │        MACON         │
 │  114.2   MCN   89    │
 └──────────────────────┘
```

Explain the frequencies above the shadow box.
A. In addition to the four frequencies normally available at all FSSs, this particular FSS uses the frequencies 122.4 and 122.1. This FSS can receive and transmit on 122.4 but can only receive on 122.1.

Q. Given the following communication box (not shadow box):

```
           122.1R
 ┌──────────────────────┐
 │        VIENNA        │
 │  116.5   VNA   112   │
 └──────────────────────┘
      └─ MACON ─┘
```

Your destination is a non-military airfield in the vicinity of this VOR. You want to close your flight plan by radio just prior to landing. Explain how you would do this.
A. In the previous example the shadow box indicated that the MACON VOR had an associated Flight Service Station (MACON RADIO). Flight Service Stations are often responsible for a large area of coverage. In order to ensure adequate reception by participating aircraft, FSSs make use of VORs (as transmitters) for their area. In the example above, MACON RADIO is the controlling FSS. You can contact them on 122.1. The "R" indicates that they can only receive on this frequency so you'll need to "listen in" on the VIENNA VOR, 116.5, in order to hear a response (ie, "Macon Radio, Army 12345, VIENNA"). Of course, you can still attempt communication on 122.2 or 255.4 since these frequencies are always available; however, in this case tell flight service what frequency you are transmitting on so they can respond appropriately and keep in mind that altitude and terrain may affect reception.

NOTE: In some cases, a frequency above a communication box or shadow box will be followed by a "T". This indicates that the FSS can only transmit on that frequency. Again, make sure that you tell flight service what frequency (or navaid) you are listening on.

Q. Now that you know all of the frequencies on which to contact Flight Service,

how can you contact them by telephone?
A. Dial 1-800-WX BRIEF (992-7433).

Q. On the ELA chart VOR airways are depicted in _________ (color) and NDB airways are depicted in __________ (color).
A. Black; brown.

NOTE: Some NDB airways are depicted and should not be confused with Military Training Routes (MTR), which are depicted in a lighter shade of brown.

Q. Victor airways contain certain minimum items of information, those being an airway name or label, a direction (magnetic course), distance, and a Minimum Enroute Altitude (MEA). Explain the purpose of an MEA and describe what other altitudes may be associated with a victor airway.
A. A Minimum Enroute Altitude (MEA) is the lowest published altitude between radio fixes which assures acceptable navigational signal coverage and meets obstacle clearance requirements between those fixes.

A Minimum Obstacle Clearance Altitude (MOCA) is the lowest published altitude in effect between radio fixes on a victor airway or route segment of a victor airway which meets obstacle clearance requirements for the entire route segment and which <u>assures acceptable navigational signal coverage only within 25 SM (22NM) of a VOR</u>.

A Minimum Reception Altitude (MRA) is the lowest altitude at which an intersection can be determined. This is indicated on the airway by an "R" enclosed in a "flag."

A Minimum Crossing Altitude (MCA) is the lowest altitude at certain fixes at which an aircraft must cross when proceeding in the direction of a higher MEA (must be at or above the designated MCA prior to "crossing" the intersection or fix in order to meet MEA requirements). This is indicated on the airway by an "X" enclosed in a "flag."

A Maximum Authorized Altitude (MAA) is a published altitude representing a maximum usable altitude for an airspace structure or route segment. Above this altitude adequate reception of navaid signals is not assured.

Q. Explain the semi-circular rule.
A. IFR: for IFR flights in controlled airspace, use an odd thousand altitude if your heading is between 0° and 179° (magnetic); use an even thousand altitude if your heading is between 180° and 359° (magnetic).

VFR: for VFR flights above 3,000′ agl, use an odd thousand altitude plus 500′ if your heading is between 0° and 179° (magnetic); use an even thousand altitude plus 500′ if your heading is between 180° and 359° (magnetic).

MEMORY AID: NEODD - north thru east = odd thousand (plus 500′ for VFR)
SWEVEN - south thru west = even thousand (plus 500′ for VFR)

ENROUTE LOW ALTITUDE (ELA) CHARTS
and
STANDARD INSTRUMENT DEPARTURES (SIDs)

Q. Given the following victor airway:

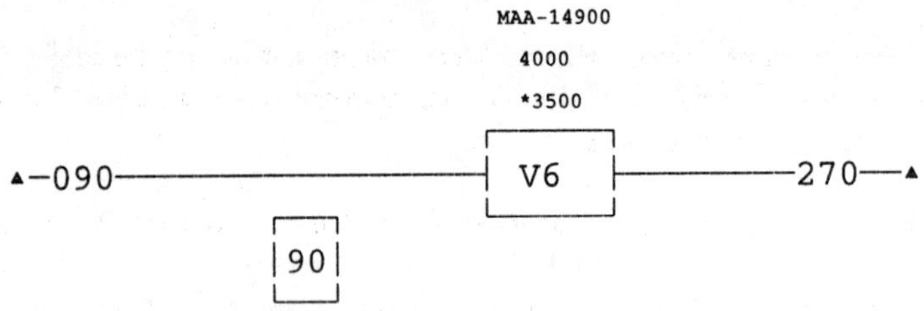

What is the name of the airway?
If you are east-bound on this airway, what is the <u>minimum</u> altitude you should file for? West-bound?
What is the maximum altitude for which I will be assured of acceptable navigational coverage for the route segment?
What is the distance between compulsory reporting points?
How are non-compulsory reporting points designated on an airway?
How are intermediate distances or distances between non-compulsory reporting points depicted?
A. The name of the airway is "V6".
Eastbound IFR = 5,000'; eastbound VFR = 3,500' (remember however, that while 3500' (MOCA) assures obstacle clearance, it only assures acceptable navigational signal reception within 22 NM of a VOR. Hence, 5,500' may be a better option).
Westbound IFR = 4,000'; westbound VFR = 4,500'.

NOTE: Keep in mind that the semi-circular rule only applies to VFR traffic above 3,000' agl. So in this example, if the terrain were at 1,000' msl then VFR traffic would not be subject to the semi-circular rule until 4,000' msl.

Maximum Authorized Altitude = 14,900'
90 miles between compulsory "▲" reporting points (and/or navaids).
Non-compulsory reporting points are designated by an open triangle (not filled in black or brown).
Intermediate distances between non-compulsory reporting points are designated numerically but are not boxed in.

Q. You are flying from ABC to DEF along the victor airway illustrated below. What victor airway should you file for?

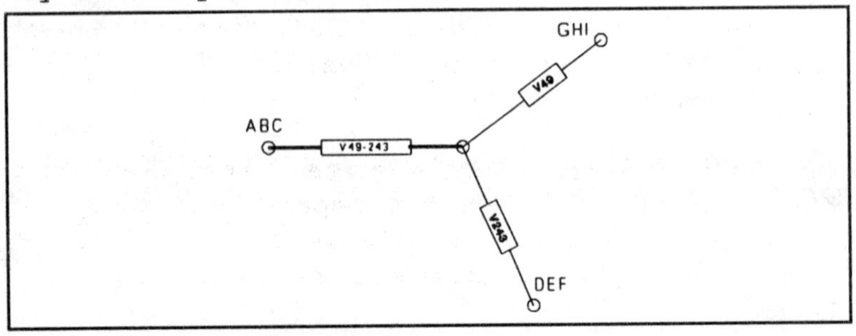

ENROUTE LOW ALTITUDE (ELA) CHARTS
and
STANDARD INSTRUMENT DEPARTURES (SIDs)

A. You should file for "V243". If GHI were your destination then you would file for "V49".

Q. At what altitude does the semi-circular rule take effect?
A. For IFR flights - when within controlled airspace (normally above 1200'agl).
For VFR flights - when above 3000' agl.

Q. When traveling on a victor airway between VORs, when should you change VOR frequencies?
A. The frequency changeover point (COP) is considered to be one of the following:
- Midway between the navigational facility for straight route segments;
- At the intersection of a radial or course forming a dog leg route segment;
- At the published COP location (see question below).

Q. Describe the meaning of the following symbol on a victor airway:

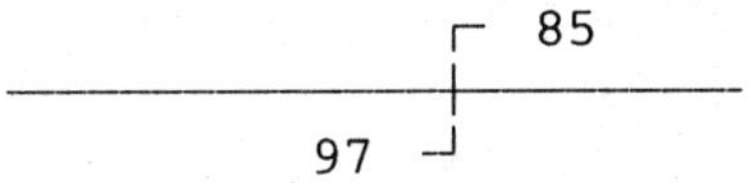

A. This symbol represents a VOR changeover point on the victor airway. It is used when the VOR changeover should occur at other than the midpoint between stations. The respective mileages to the VORs are provided to aid in determining the location of the changeover point.

Q. What is indicated by the following symbology on the EL chart?

A. This is used on a victor airway to indicate an MEA change.

Q. What is meant by a Minimum IFR Altitude (MIA)?
A. The minimum altitude for IFR operations as prescribed for airways, route segments, and instrument approach procedures. If no applicable minimum altitude is prescribed (off airways, direct routes, etc), then the following minimum IFR altitude rules apply:

(1) In designated mountainous areas (see AIM), 2000 feet above the highest obstacle (AHO) within a horizontal distance of 4 SM from the course to be flown; or

(2) In other than mountainous areas, 1000 feet AHO within a horizontal distance of 4 SM from the course to be flown; or

(3) As otherwise authorized by ATC.

Q. On most ELA charts the background is white. What does this represent?
A. Unless otherwise indicated, Class E airspace. In some areas of the United States, Class G (uncontrolled) airspace exists above 1200' AGL and, in fact, extends all the way up to 18,000'. In areas where this is the case the ELA chart will be shaded brown.

ENROUTE LOW ALTITUDE (ELA) CHARTS
and
STANDARD INSTRUMENT DEPARTURES (SIDs)

Q. How is Class B airspace indicated on the ELA chart? Class C airspace?
A. By a shaded blue area enclosed by a solid blue line; by a shaded blue area enclosed by a dashed blue line.

Q. What is represented by alternating blue and white "stripes" at some airfields on the ELA chart?
A. This represents the Mode C area surrounding Class B airspace.

Q. What is the purpose of a Military Training Route (MTR)?
A. The maximum indicated airspeed for all aircraft operating below 10,000' msl is 250 kts. Specific areas of defined vertical and lateral dimensions have been approved and designated for the conduct of military flight training at airspeeds in excess of 250 kias.

NOTE: Military aircraft operating on Military Training Routes pose a significant hazard to VFR traffic. Extra vigilance should be exercised when crossing these areas.

Q. MTRs will be preceded by the letters "IR" or "VR". What do these letters indicate?
A. IR - IFR Training Route; VR - VFR Training Route.

Q. MTRs are additionally identified by a three or four number character. What is the difference between these?
A. MTRs identified by three number characters have one or more segments above 1500' agl (ie, IR-206, VR-207); those MTRs with no segments above 1500' agl are identified by four number characters (ie, IR-1206, VR-1207).

Q. What is an ARTCC?
A. Air Route Traffic Control Center (pronounced ART'-SEE); it is a facility established to provide ATC service to aircraft operating on IFR flight plans within controlled airspace and principally during the enroute phase of flight. Time and workload permitting, certain advisory/assistance services may be provided to VFR aircraft.

Q. What does the following symbol on the ELA chart indicate?

A. The boundary between two ARTCCs.

Q. On several areas of the ELA chart you will find what looks like light blue postage stamps with information inside. What are they?
A. If, for instance, you are within the MEMPHIS ARTCC, then Memphis Center will be responsible for monitoring your progress on radar and handing you off to the appropriate approach control, or another Center, etc. In order to efficiently control the large number of aircraft within an ARTCC at any one time, the area is divided into sectors with one or more controllers working each sector. These

"postage stamp" looking symbols represent the discrete VHF and UHF frequencies for these sectors. You can expect on initial contact to use one of the "postage stamp" frequencies closest to your location.

Q. What does a light green dotted line indicate on the ELA chart?
A. Time zones.

Q. All mileages on the ELA chart are ____________ (statute/nautical).
A. Nautical miles unless otherwise noted.

Q. All altitudes on the ELA chart are _________ (msl/agl).
A. MSL, unless otherwise noted.

Q. What does the symbol "x" on an airway denote?
A. Mileage breakdown.

Q. All navaid symbols show a solid line with a half-arrowhead at the top. What does this represent?
A. Magnetic north.

Q. What is indicated by an area on the chart surrounded by a square composed of dashed grey lines?
A. This indicates an area of enlargement. AREA CHARTS are available for through-flights.

Q. An accurate intersection is made up of radials which intersect at an angle of at least _______°.
A. 30 (radials intersecting at 90° would form the most accurate intersection).

Q. When several radials come together at the same intersection, what is used to denote which radials have been FAA checked for accuracy and should be used to fix the intersection?
A. Small arrows.

Q. What does SID stand for and what is its purpose?
A. Standard Instrument Departure; the purpose of a SID is to expedite and standardize the flow of traffic around certain congested airports.

Q. How can I find out if an airport has a SID?
A. If the airport is blue on the ELA chart you can expect to find a SID (if it has one) in the DOD Terminal FLIP. This FLIP is published in two volumes (East/West), is bound at the top, and has a black band at the top and bottom of each volume. Low Altitude Instrument Approach Procedures booklets (approach plates) <u>may</u> contain some SIDs. If the airport is green on the ELA chart a SID could be found in the U.S. GOV'T Terminal Procedures booklet (similar to DOD FLIPs but have two black bands near the top and a loose-leaf binding). The Index of Instrument Approach Charts in the front of the FLIPs will give a listing of

all approaches and departures for each airport.

Q. After a aircraft is established on a SID and subsequently is radar vectored or cleared off the SID/SID transition, pilots shall consider the SID canceled unless the controller says what?
A. "Expect to resume SID."

DEPARTMENT OF DEFENSE (DOD)
FLIGHT INFORMATION PUBLICATION (FLIP)

The preponderance of information contained in FAA and DOD flight information publications regarding IFR flight procedures alone can seem overwhelming to a Flight School student. The important rule at this stage, and indeed throughout much of your aviation career, will not be to have it all "memorized" but to know where to look to obtain the necessary information.

The following questions represent a relative cross-section of the types of questions that may be asked on instrument exams. These questions are presented in order to show the "logic" used in finding answers.

Q. Where can you look to find information concerning the NOTAM system?
A. Look in the GP, "Index of Aeronautical Information", under NOTAM SYSTEM and it will direct you to chapter 5 of the GP.

Q. While checking the NOTAMS board in Operations you see a printout that looks something like this:

A) KOZR
B) WIE
C) UFN
E) QCPAU

What is this? Where would you look to find out what it means?
A. This is a Class I NOTAM. More often than not there will also be a textual explanation of the coded NOTAM. If this is not provided then it must be decoded.

Look again in the GP, "Index of Aeronautical Information", under NOTAMS. Of the four entries dealing with NOTAMS, "NOTAM CODE" would be the logical choice. Proceed as directed to the FIH. Notice from the front cover that Section F deals with "FLIP and NOTAM Abbreviations/Code". Page F-1, the Table of Contents, refers you to Page F-23. This section describes what each of the above lines represents.

Q. While performing an airborne VOR check at Stonyfork, PA, what altitude should you be at?
A. In the GP, "Index of Aeronautical Information", under VOR RECEIVER CHECKPOINTS you'll see that the ES(AK) and the AP/1,2,3,4 all contain this information. The term "conterminous" (see GP, "Explanation of Terms") refers to the 48 adjoining states and the District of Columbia so rule out the Enroute Supplement for Alaska. You'll need the AP/1. AP/2, 3, and 4 are for other areas of the world (see the chart on back of the GP). Looking in the AP/1 Table of Contents, you'll see that Chapter 3 deals with North America. The United States begins on page 3-9 and further along you can see VOR RECEIVER CHECKPOINTS beginning on page 3-146. Starting on page 3-146, scan alphabetically for Stonyfork, PA. Here you will see that you need to be at 3600'.

Q. What is a Radio Class Code? Where could you look to determine what these codes are?
A. All navigational aids are required to have a designated Radio Class Code which gives the reception limits usable for random/unpublished navigation.

Start at the "Index for Aeronautical Information" in the front of the GP. You will not find "Radio Class Codes" listed but you will see a listing for "Radio Beacon Data". Here you see that this information can be found in the Enroute

(IFR) Supplement. Thumbing through the legend information in the front of the ES you will find Radio Class Codes at item 31.

Q. What are the four operational levels of AWOS?
A. If you know what AWOS stands for then you are ahead of the game, but suppose you don't - time to play detective. Probably the best place to look for abbreviations is Section F of the Flight Information Handbook (FIH). Look on the front of the FIH, it breaks everything down for you.

In Section F you'll see that AWOS stands for Automated Weather Observation System. Now you can do several things. Try looking up the definition under "Explanation of Terms" in the GP. Hmmm, just a definition, nothing about operational levels. If you are still stumped try looking in the "Index for Aeronautical Information", also in the GP. Here you'll see that it refers you to the FIH.

Looking on the front of the FIH you'll see that there is an entire section devoted to meteorological information, Section C (Hmmm, could have saved a step by noticing that earlier!). Go to page C-1, the Table of Contents for Section C. The first entry is AWOS and it tells you to refer to page C-33. At the bottom of this page you'll find the four operational levels of AWOS.

Q. You look under a particular airfield and see the following **COMMUNICATIONS** data: "**CTAF/UNICOM** - 123.3 **PTD** - 372.2". What is the UHF frequency used for?
A. Since this information comes from the IFR Supplement, the logical place to check is the A/FD legend in the front of the book. If you look at the example (HOG HOLLER INTL) you'll find "PTD" under the communications data and you are referenced to paragraph/section 27. If you look there you won't find anything. Next try looking in the abbreviations section (Section F) of the FIH. Here you'll see that "PTD" stands for "Pilot to Dispatcher." You could have also found "PTD" in the abbreviations section of the ES immediately following the A/FD legend.

Q. Where would you look to find information on Special Use Airspace?
A. This is thrown in to give you a typical example of "Armese" - you know, that special language that the Army speaks?!! If you look in the GP, "Index for Aeronautical Information", under "Special Use Airspace" you will come up blank. However, if you look under "AIRSPACE, SPECIAL USE", BINGO! refer to AP/1A, 2A, etc. Just a little "INFORMATION, TIDBIT OF".

Q. Where can I look to find information on vortex avoidance procedures involving wake turbulence?
A. This one is a little tougher. Look in the GP, "Index for Aeronautical Information". WAKE TURBULENCE - nothing. TURBULENCE, WAKE - nothing. VORTEX - nothing. Next look under "Explanation of Terms" in the GP. There's a definition but nothing about vortex avoidance procedures (if you see "Jet Blast" and "Acft Classes" like it suggests you'll still come up empty).

Hmmm. Try the "Index of Subjects" in the front of the FIH. Still nothing. Wait a minute. This could be considered an emergency procedure, right? Look under Section A of the FIH. Again, nothing.

DEPARTMENT OF DEFENSE (DOD)
FLIGHT INFORMATION PUBLICATION (FLIP)

Let's see - don't give up yet. Scan the table of contents in the GP - FLIP Program, Flight Plans, Pilot Procedures, wait a minute, what's this "Supplementary Information"? BINGO! AIRCRAFT WAKE TURBULENCE Page 5-11. Scanning through that section you'll find, on page 5-12, Vortex Avoidance procedures.

Q. Where in DOD FLIPs could you look to find the Special VFR (SVFR) minimums for Cairns AAF?

A. This one would be pretty tough if you didn't already know where to start looking. The normal procedure of looking in an index or table of contents will not reveal anything. Of course, the information would be found in the SOP but that's not what we're talking about.

The clue here is that the question asks about a specific geographic area, specifically Cairns AAF. Check the Area Planning document (AP). The obvious choice here would be the AP/1 since we're dealing with NORTH AND SOUTH AMERICA. In the Table of Contents you'll see Chapter 3 - NORTH AMERICA. Well, that narrows it down a little. Scanning further you'll see the section on the UNITED STATES begins on page 3-9. The short remark on VISUAL FLIGHT RULES won't help but alas we get to SUPPLEMENTARY AIRPORT REMARKS. Look in this section under Cairns, AAF, and you'll find the SVFR minimums applicable only to Fort Rucker operated airfields/heliports.

Q. You see the following line in the IFR Supplement under an airfield:

"**REMARKS** - Attended 1100 - 0300Z ++"

What is meant by the "++" symbol?

A. Again, since you are using the IFR Supplement, the first logical step is to check the legend in the front of the book. Scanning the example (HOG HOLLER INTL) for the "++" symbol you will find it under **REMARKS** here as well. Referring to section 26 won't reveal anything. If, however, you begin scanning at the beginning of the legend you will find it under **TIME CONVERSIONS** at section 11. Here you will see that the symbol "++" following operating hours indicates that, during periods of daylight saving time, the effective hours will be one hour earlier than shown.

The following question should test your ability to find a variety of subjects.

Q. Given the following DOD FLIPs - General Planning (GP), Area Planning (AP), Enroute IFR Supplement (ES), Enroute Low Altitude charts (ELA), Flight Information Handbook (FIH), and Terminal Charts (TC - approach plates), state where you would look for the following information/procedure:

a. Two-way radio failure
b. Services available at a specific airport
c. Time zones of the world
d. Designated mountainous areas
e. How to fill out a flight plan
f. The explanation of a term you do not understand (definitions)
g. An abbreviation you do not understand
h. A chart symbol you do not understand

---

i. Airport coordinates
j. About weather info (AWOS, ATIS, EFAS, PMSV, PIREPs, etc)
k. Information regarding the NOTAM system
l. Use of pilot-controlled airport lighting
m. Recommended procedures for any emergency (uncertainty, distress, lost)
n. ATIS frequencies
o. Airfield elevations
p. Prohibited, restricted, and warning areas
q. Transponder procedures
r. ATC light signals
s. Position reporting requirements
t. Diagram for destination airport
u. Aircraft approach categories
v. Change of flight plan procedures
w. Aircraft classes
x. FAA Air Traffic Control procedures
y. VIP codes
z. Don't know where to look

A. a. FIH, section A
b. ES
c. FIH, section E
d. AIM*
e. GP, chapter 4
f. GP, chapter 2
g. FIH, section F; ES, section A
h. ELA legend or TC legend
i. ES
j. FIH, section C
k. GP, chpt 5; FIH Section F
l. FIH, section B
m. FIH, section A
n. ES, ELA, TC
o. ES, ELA, TC
p. ELA, AP
q. GP, chapter 5
r. FIH, section A
s. FIH, section B; ES, back cover
t. TC
u. GP, chapter 2
v. ES, back cover
w. GP, chapter 2
x. FIH, section B
y. GP, chapter 4
z. Start with the "Index of Aeronautical Information" in the front of the GP. Also try the FIH, "Index of Subjects".

* Designated mountainous areas used to be published in the AP and is presented here for this reason. Such areas may now be found in the Airman's Information Manual (AIM). Look in the AIM index under "Designated Mountainous Areas" for the appropriate location in the AIM.

As mentioned at the beginning of this chapter it will be most important for the flight school student to be able to <u>find</u> information. In fact, most of your instrument exams will be open-book to the extent that you will have access to numerous publications (DOD FLIPs).

However, as your aviation career progresses you will want to have more and more of this information tucked neatly away in memory. Refer to Chapter 10 of this study guide for detailed questions concerning general and instrument flight rules.

FLIGHT PLANS

Q. Where in the DOD FLIPs will you find information regarding how to properly fill out a flight plan?
A. GP, Chpt 4.

Q. What is the primary purpose of a flight plan?
A. To relay the desires of the pilot to the controller.

Q. What is a DD Form 175?
A. Military Flight Plan

Q. What is a DD Form 175-1?
A. Flight Weather Briefing.

Q. What is an FAA Form 7233-1?
A. Civilian (FAA) Flight Plan.

Q. When must a DD Form 175 be completed?
A. For all military flights (VFR & IFR) departing US installations having a military base operations facility.

NOTE: Some installations have defined operating procedures that do not require a DD Form 175 for local flights as long as certain minimum flight-following requirements can be met. In these instances, a flight "strip" will be on file with unit operations and an operations log completed.

Q. When would a military pilot use an FAA Form 7233-1?
A. This form may be used in lieu of a DD Form 175 when departing US installations not having a military base operations facility.

Q. The first block on DD Form 175 is the date. Dates can be either zulu or local just like time. Which one should be entered here?
A. Date of flight in <u>local</u> time.

Q. The aircraft call sign entry is limited to how many characters?
A. 7

Q. The aircraft call sign is normally a "code" followed by the last five digits of the aircraft tail number. What will this "code" be for most Army aircraft?
A. R

Q. What would be the "code" for helicopters performing MEDEVAC missions?
A. E (<u>E</u>VAC).

Q. What should be entered in the AIRCRAFT DESIGNATION AND TD CODE block for a Huey helicopter equipped with a Mode C capable transponder.
A. UH-1/U (when saying this aloud, correct terminology would be "UH-1 slant Uniform).

Q. What would be entered in the AIRCRAFT DESIGNATION AND TD CODE block for a

flight of four Blackhawk helicopters?
A. 4/UH-60/U.

Q. What does the TD suffix "X" indicate?
A. No transponder.

Q. In the TYPE FLT PLAN block, the letter ______ would be entered for an IFR flight and the letter _____ for a VFR flight.
A. I; V.

Q. Can IFR and VFR route segments be combined on the same flight plan?
A. Yes.

Q. Can IFR and VFR route segments be combined on the same line of the flight plan?
A. No.

Q. When should a "D" be entered in the TYPE FLT PLAN block?
A. For flights penetrating or operating within the ADIZ (Air Defense Identification Zone).

Q. For IFR flights penetrating or operating within the ADIZ, should a "D" be entered in the TYPE FLT PLAN block?
A. No, an "I" should be entered as normal.

Q. Where can you find information regarding ADIZ procedures?
A. IFR Enroute Supplement, Section C.

Q. What figure should be entered in the TRUE AIRSPEED block?
A. The TAS to be maintained at <u>initial</u> cruising altitude (refer to appropriate Performance Planning charts).

Q. When is a TAS entry <u>not</u> required?
A. Only for local VFR flights.

Q. One of three different entries may be made in the POINT OF DEP block. What are they?
A. - The location identifier of the departure airport; or
- The point (navaid or fix) where IFR will begin; or
- The installation name if there is not a location identifier.

Q. What time should be entered in the PROPOSED DEP TIME (Z) block?
A. The proposed departure time in Coordinated Universal Time (UTC), also known as "Zulu" time.

NOTE: Keep in mind that sufficient time should be allowed for Operations to process the flight plan. <u>For IFR flights allow a minimum of 30 minutes for processing.</u>

FLIGHT PLANS

Q. If there is an unusual delay and you cannot meet your proposed departure time, what should you do?
A. Normally, your flight plan will be discarded if you have not departed within one hour of the filed proposed departure time. If your delay is expected to be longer than this then you should advise Base Operations or Flight Service and provide an updated proposed departure time.

Q. When departing non-military airports, what can be the effect of not reporting departure time to the tie-in FSS?
A. You may arrive at your destination unannounced.

Q. What altitude should be entered in the ALTITUDE block?
A. The initial cruising altitude.

Q. How should the altitude be entered in the ALTITUDE block?
A. In hundreds of feet (ie, 4000' = 40, etc).

Q. How would you signify that you wish to change altitudes during subsequent route segments?
A. Enter the requested altitude and location of the change in the "REMARKS" block.

Q. What would the letters "OTP" in the ALTITUDE block indicate?
A. That you wish to fly IFR/VFR-on-top. An altitude may also be specified if desired (ie, "OTP 125").

Q. Your IP hands you a flight plan with "80B100" in the ALTITUDE block. What does that mean?
A. Your IP is requesting a block altitude of between 8,000 and 10,000 feet.

Q. If a radar departure or VFR climb is desired, where should this be indicated?
A. In the "REMARKS" section.

Q. What should be the first point in the ROUTE OF FLIGHT section?
A. The planned navaid or fix for entering the enroute structure (ie, VOR, NDB, named intersection, etc) for IFR flights. For VFR flights the ROUTE OF FLIGHT section should contain entries that most clearly define the route of flight. The absence of an entry indicates DIRECT flight.

Q. If a SID is used in the ROUTE OF FLIGHT section, what information should be included with it?
A. The SID identification followed by either the SID termination point or the transition fix (ie, Junction-Five Dep, enter "JCT5.JCT", or Junction-Five with Abilene transition, enter "JCT5.ABI").

Q. The ROUTE OF FLIGHT section should be used to "clearly define the route of flight." What does the absence of additional information between navaids or fixes in the route of flight indicate?
A. This indicates direct flight (ie, ABB OOM).

FLIGHT PLANS

Q. How would a transition from one airway to another at an unnamed intersection be entered on the DD Form 175?
A. Enter the designations of the two airways, separated by a space (ie, YKM V4 V187 TCM).

Q. If you use an L-class nondirectional beacon as a navaid it will have a two-letter identifier. Can you file using this navaid?
A. Yes. However, ARTCC computers will not accept two-letter identifiers or identifiers with more than five letters. All L-class beacons will have a five-letter name; use this in the ROUTE OF FLIGHT section.

Q. All intersection names consist of _____ letters.
A. Five

Q. What should be the last point entered in the ROUTE OF FLIGHT section for a VFR flight?
A. The point from which the final leg is begun to the destination.

Q. What should be the last point entered in the ROUTE OF FLIGHT section for an IFR flight?
A. The identification of the nearest IAF, navaid, first point of intended landing, or published fix which most clearly establishes the route of flight to the destination.

NOTE: If a Standard Terminal Arrival (STAR) is used, the last point should be the STAR identifier, placed after the transition fix.

Q. What is required to be entered on all successive legs of a stopover flight plan?
A. 1.) The amount of fuel on board (HH+MM): should be in parenthesis following the last entry of each leg.
2.) If an alternate is required: the airport identifier and the ETE to the alternate in parenthesis with the fuel on board entry (ie, 2+07 OZR 0+30).

FUEL ALT/ETE

Q. What is the purpose of a terminal area delay and how would it be entered on the flight plan?
A. One reason for the terminal area delay is to provide pilots with the option of conducting practice approaches or traffic pattern work at an airport that is in the route of flight without having to close and re-file or modify the flight plan.
- The identification of the delay location airport should be entered as the last item in the route of flight, <u>without</u> an entry in the TO block.
- Estimate the amount of time to the delay airport and enter in the ETE block.
- On the next line enter an "R" with a circle around it (to indicate that this should be transmitted as a remark). After the "R" enter a "D" (for delay), followed by the time of the delay (HH+MM), the identifier of the delay airport,

---

and the airport of the final destination (ie, (R) D 0+15 BSM RND).

Q. Flight plans are normally presented as one continuous flight. However, if a holding delay is desired, how should it be entered?
A. Enter the holding delay immediately following the holding fix.
Without delay: (DUMPS J17 J86 AUS BSM)
With 10-minute holding delay: (DUMPS/D 0+10 J17 J86 AUS BSM)

Q. When should "LOCAL" be entered in the ROUTE OF FLIGHT section?
A. If the flight is to be conducted VFR within the designated local flying area and expected to terminate at the base of departure or an installation under the operational control of the base of departure.

Q. What should be entered in the TO column of the flight plan?
A. The location identifier of a full stop or final destination airport. The identifier should be entered opposite the last line entry in the ROUTE OF FLIGHT section. If there is no location identifier, enter the installation name.

Q. How should the ETE be calculated for a VFR flight plan?
A. The time from takeoff to a position over the destination airport, <u>including</u> known or preplanned enroute delays.

Q. How should the ETE be calculated for an IFR flight plan?
A. The time from takeoff or departure to the last fix shown in the ROUTE OF FLIGHT section, <u>excluding</u> known or preplanned enroute delays.

REMEMBER: ETE for VFR flight plans should always include known enroute delays; ETE for IFR flight plans should not.

Q. The REMARKS section should be used to enter information essential to safe and efficient control of air traffic. What should be entered in the REMARKS section for an aircraft on an IFR flight plan equipped with ADF as the only enroute navaid?
A. "ADF ONLY."

Q. Explain the use of the following abbreviations in the REMARKS block:

PPR 4306
MGM-S
DHN-R

A. "PPR" stands for Prior Permission Required. A PPR may be required to land at a certain airport, or a PPR may be required for fuel. Whatever the case, enter the PPR number so that it will be available to the appropriate facility and minimize confusion.

"S" stands for Services. This indicates that service (normally refuel) will be required at MGM.

"R" stands for Remain over night. This indicates that you will be remaining overnight at DHN.

FLIGHT PLANS

Q. What other items of information may or should be included in the REMARKS section?
A. - Any other pertinent information deemed necessary to be transmitted to the destination.
   - Void Time.

Q. When is a VOID time required to be entered in the REMARKS section?
A. For all stopover and/or terminal area delay flight plans.

Q. How is VOID time calculated?
A. Calculate the total time from takeoff to final destination and round to the next whole hour (ie, for total time = 1+20, enter "VOID 2+00).

Q. It is the responsibility of each pilot in command transporting VIPs to insure that the flight plan reflects the high rank status of such individuals. The appropriate entry in the RANK AND HONOR CODE section is essential to preclude any embarrassment to the VIP, the airfield commander, or the aircraft commander. What code would be entered for an Army Colonel (with no honors on arrival)? An Army Major-General requesting informal visit with the commander?
A. R7O; R5N (see GP chapter 4, section 4-3).

Q. What other information is required in the RANK AND HONOR CODE section for flight plans with multiple legs (stopovers)?
A. Enter the pick-up and drop-off points of the VIP (ie, R5O BLV to LRF).

Q. How should the FUEL ON BOARD be calculated?
A. Enter the total time that the aircraft can stay aloft while flying the planned profile with the fuel available at initial takeoff (use performance planning data).

Q. What should be entered in the ALTERNATE AIRFIELD block?
A. - If an alternate is required, enter the location identifier (or station name if no location identifier).
   - If IFR on a stopover flight plan, the alternate for the first point of intended landing is entered here.
   - Alternates for subsequent stops, if required, are entered in the ROUTE OF FLIGHT section.
   - If no alternate is required, leave blank.

Q. How should the ETE to alternate be calculated?
A. Calculate the time from the missed approach point to the missed approach holding point, to include one circuit in the holding pattern. Add to that the time required from the missed approach holding point to the alternate airfield, to include time for approach and landing (FM 1-240, App B).

Q. What information should be entered in the NOTAMS/WEATHER blocks?
A. These blocks are included as pre-flight reminders and may be used as directed locally. In the absence of local directives it is advisable to enter the local time that NOTAMS were checked in the NOTAMS block, and the initials of the

weather briefer in the WEATHER block.

Q. When filing a flight plan the pilot in command must ensure that a DD Form 365-4 is file with the flight plan or certify on the flight plan that the loading for the proposed flight does not exceed the loading limits. Knowing this, what should be entered in the WEIGHT AND BALANCE block?
A. Either:
- "Attached", if the DD Form 365-4 is attached; or
- (Filed at)_____/(date)_____, when citing a previously filed DD Form 365-4 for a normal loading (ie, OZR/2 JUN 86).

Q. Whose name should appear in the SIGNATURE OF APPROVAL AUTHORITY block?
A. The pilot-in-command of the aircraft.

Q. If the list of crew members will not fit in the given blocks on the DD Form 175, where should they be listed?
A. The crew list may be continued on the back of the DD Form 175 original copy, or attached on a separate piece of paper. The attachment may be a copy of crew orders, or any piece of paper with the appropriate information on it. When a crew list is attached, check the appropriate box.

Q. Are Social Security Numbers required to be entered in the SSN column at the bottom of the flight plan?
A. No. Disclosure under the Privacy Act is voluntary but failure to provide the SSN could result in denial of flight plan processing.

Q. For formation flights, where should the appropriate aircraft tail numbers be entered?
A. Under the ORGANIZATION AND LOCATION column (attach sheets of paper as necessary or include information on back of the original).

Q. Where on the flight plan would you indicate that you were carrying hazardous cargo?
A. In the REMARKS section (ie, "Hazardous Cargo").

Q. What are the two main IFR navigation systems used by Army aviators?
A. NDBs and VORs.

Q. What does VOR stand for?
A. Very High Frequency (VHF) Omnidirectional Range.

Q. What are the two basic components of a VOR?
A. A VOR transmitter (on the ground) and the receiver (in the aircraft).

Q. What are the VHF frequency ranges for VOR stations?
A. 108.0 - 112.0 (even tenths)
112.0 - 118.0 (all tenths)

NOTE: Odd tenths in the VHF frequency range 108.0 - 112.0 are reserved for localizer frequencies.

Q. How does a VOR work?
A. The VOR transmitter emits two separate signals on the same frequency - the reference phase signal and a variable signal. The reference signal remains the same. However, there are 360 slightly different or variable signals transmitted in different directions. These signals are called radials and radiate outwards from the center of the VOR like spokes from a wheel. By comparing these signals, the VOR receiver in the aircraft can indicate its position in relation to the tuned station.

Q. How do you "tune-identify" a VOR station?
A. Tune the appropriate frequency from VFR/IFR charts into the VOR receiver and check that the station identification (dots and dashes) corresponds with that station. Also check that an indication is received from the appropriate instruments.

Q. What are the primary and secondary instruments used to help navigate using VORs?
A. Primary - Course Deviation Indicator (CDI) with OmniBearing Selector (OBS).
Secondary - The No. 2 needle on the Radio Magnetic Indicator (RMI).

Q. Using the RMI, how do you determine the magnetic course to the VOR?
A. Look under the head of the No. 2 needle.

Q. Using the RMI, how do you determine what radial from the VOR you are on?
A. Look under the tail of the No. 2 needle.

Q. What is the accuracy of course alignment for a VOR?
A. Generally, +/- 1°.

Q. Is VOR reception limited to line-of-sight?
A. Yes.

Q. Does atmospheric static interfere with tuning and identifying a VOR signal?

A. No.

Q. Name some ways that VORs are used for navigation.
A. Victor airways, VOR instrument approaches, position fixes, and navigation for direct routes of flight.

Q. What is a Victor airway?
A. A Victor airway is a "highway" or corridor in the sky between two VORs. It is situated in controlled airspace (officially designated as Class E airspace) between 1200' AGL and 18,000' AGL. They are depicted on VFR and IFR charts. Such factors as obstacle clearance and reception are guaranteed if flown at or above the published altitudes for the airway.

Q. What is the normal width of a Victor airway?
A. 4 NM either side of centerline.

Q. North/south victor airways are _______(odd/even) numbered and east/west airways are always ______ (odd/even) numbered.
A. Odd; even (just like interstates on the ground).

NOTE: This reflects the "general" direction of the airway from beginning to end and not necessarily between route segments, fixes, or VORs.

Q. As a minimum, victor airways will always provide what four items of information?
A. The airway number, a direction (magnetic course), a distance, and a Minimum Enroute Altitude (MEA).

Q. What other items of information may be given concerning a particular victor airway?
A. VOR changeover points, Minimum Obstruction Clearance Altitude (MOCA), Maximum Authorized Altitude (MAA), Minimum Reception Altitude (MRA), Minimum Crossing Altitude (MCA), and compulsory/noncompulsory reporting points.

Q. What defines the reception limits of unrestricted navaids which are usable for random/unpublished navigation (or, in other words, what defines the reception limits for navaids usable for direct routes - off victor airways)?
A. The Standard Service Volume (SSV).

Q. By SSV, VORs can be classified into one of three types. What are they?
A. (1) Terminal VOR - designated by the symbol (T) on IFR charts and in the IFR Supplement. Unless otherwise noted, interference-free reception is normally anticipated at radial distances of 25 NM from the station and from approximately 1,000' AGL to 12,000' AGL.

(2) Low altitude VOR - designated by the symbol (L) in the IFR Supplement. Unless otherwise noted, interference-free reception is normally anticipated at radial distances of 40 NM from the station and from approximately 1,000' AGL to 18,000' AGL.

(3) High altitude VOR - designated by the symbol (H) in the IFR Supplement.

Normal anticipated interference-free reception (unless otherwise noted):

below 14,500'........ 40 NM
14,500' to 17,999'... 100 NM
18,000' to 45,000'... 130 NM
above 45,000'........ 100 NM

NOTE: See Radio Class Codes in IFR Supplement.

Q. The answers to the above question each contain the phrase "Unless otherwise noted..." in regards to interference-free reception. What is an example of this?
A. Check the IFR Supplement for the navaid in question. You may, for example, see a phrase similar to this: "AGS VOR unuse 015°-070° byd 25 NM blw 4500'." In this case, even though AGS is a high altitude VOR (normally good for 40 NM below 14,500') it is unusable for reliable navigation between 015° and 070° beyond 25 NM if below 4500'. This could be due to a variety of factors, including terrain. It should never be assumed that interference-free reception will be received at all altitudes and distances within its classification - CHECK THE IFR SUPPLEMENT!

Q. Do SSV limitations apply to published IFR routes or procedures?
A. No, but remember that reception is only guaranteed if at or above published altitudes (not to exceed MAA).

Q. What does TACAN stand for and what does it provide?
A. Tactical Air Navigation Equipment; provides distance measuring capability for aircraft equipped with Distance Measuring Equipment (DME).

Q. What is a VORTAC and how is it used by Army helicopters?
A. A VORTAC is a VOR that is collocated with a TACAN; most Army helicopters do not have DME and thus, only the VOR portion of the VORTAC is used.

Q. What is an NDB?
A. Nondirectional Radio Beacon.

Q. What are the two basic components of an NDB?
A. An NDB transmitter (on the ground) and an Automatic Direction Finding (ADF) receiver (in the aircraft).

Q. What are the frequency ranges for NDBs?
A. 30-300 kHz for low frequency transmitters and 300-3,000 kHz for medium frequency transmitters.

Q. How does an NDB work?
A. Unlike a VOR transmitter, an NDB transmitter emits the same signal in all directions (nondirectional). A properly tuned ADF receiver will cause the No. 1 needle on the RMI to indicate the direction to the station.

NOTE: The operation of the No. 1 needle as presented here is for the RMI used in the UH-1. Navigational instruments in other aircraft may provide different indications. Check the operator's manual.

Q. Using the RMI, how do you determine the magnetic course to an NDB?
A. Look under the head of the No. 1 needle.

Q. Using the RMI, how do you determine the magnetic bearing from an NDB?
A. Look under the tail of the No. 1 needle.

Q. Is NDB reception limited to line-of-sight?
A. No.

Q. Does atmospheric static interfere with tuning and identifying an NDB signal?
A. Yes (lightning flashes will cause the needle to point toward the center of the thunderstorm).

Q. NDBs can be classified into one of four types or classes depending on the power rating. What are they?
A. L class - power rating less than 25 watts and are reliable to approximately 15 NM. Normally used as compass locators (component of ILS system).
MH class - power rating less than 50 watts and are reliable to approximately 25 NM.
H class - power rating between 50 and 1,999 watts and are reliable to approximately 50 NM.
HH class - power rating greater than or equal to 2,000 watts and reliable to approximately 75 NM.

NOTE: See Radio Class Codes in IFR Supplement.

Q. Besides power and range characteristics, what is normally the most obvious difference between an L-class and an MH-class NDB?
A. Although technically still a "nondirectional beacon," the L-class NDB is used as a Compass Locator (a component of the ILS system).

NOTE: Any class beacon may be used as a compass locator, however, L-class beacons are normally used.

Q. How can you tell an L-class beacon from all others just by looking at a VFR or IFR chart?
A. An L-class beacon will always have a two-letter identifier. An MH-class beacon may have a two-letter identifier but normally has a three-letter identifier. H- and HH-class beacons have three-letter identifiers. In addition, any beacon having a two-letter identifier will have a name consisting of exactly five letters.

Q. Name some ways that NDBs are used for navigation.
A. NDB approaches, direct routes of flight, position fix, and as a component of ILS approaches (compass locators and locator outer markers).

Q. Can commercial broadcast stations (radio stations) be used as alternate navigational aids?
A. Yes, the ADF may be tuned to most AM radio bands. However, remember that they

can be identified only when the broadcast is interrupted for "station identification," some operate only during daylight hours, and many of the low-powered stations transmit on identical frequencies causing erratic ADF indications. As a general rule, use commercial broadcast stations as a last resort and try to use only those that are designated on enroute low altitude (ELA) charts.

Q. What is a marker beacon?
A. A radio facility capable of transmitting a signal in a vertical direction only. The signal will be received only when flying directly over the facility. Its purpose is to provide a definite radio position fix, such as on an ILS approach. ILS marker beacons have a rated power output of 3 watts or less and all marker beacons transmit on a frequency of 75 MHz. Thus, the receiver, which is preset to this frequency, will respond with the appropriate audio and/or visual indications when flying directly over the beacon.

Q. What is a locator outer marker?
A. It is a marker beacon collocated with a compass locator beacon (L-Class NDB). The outer marker normally indicates a position at which an aircraft at the appropriate altitude on a localizer course will intercept the ILS glidepath. They are usually also designated as initial approach fixes (IAF) and can be identified with continuous dashes at the rate of two dashes per second.

Q. How can you tell the difference between the outer marker and the middle or inner markers (if they exist)?
A. The outer marker can be identified by a continuous series of dashes; the middle marker by alternating dots and dashes; and the inner marker as a continuous series of dots.

OUTER MARKER (OM)	- - - - -
MIDDLE MARKER (MM)	- · - · -
INNER MARKER (IM)	· · · · · ·

Q. What are the two types of instrument approaches?
A. Precision and nonprecision.

Q. What are the two types of precision approaches?
A. (1) ILS - instrument landing system
(2) PAR - precision approach radar (a ground controlled approach).

NOTE: The Microwave Landing System (MLS) is also considered a precision approach but is not used by most Army helicopters

Q. Name seven types of nonprecision approaches.
A. (1) VOR - an approach using a VOR
(2) NDB/ADF - an approach using an NDB
(3) LOC - an approach using a localizer (ILS minus glide slope information)
(4) LOC BC - an approach using the back course of a localizer
(5) ASR - airport surveillance radar (a ground controlled approach)
(6) SDF - simplified directional facility

(7) LDA - localizer-type directional aid

Q. A Decision Height (DH) is always associated with __________ approaches, whereas a Minimum Descent Altitude (MDA) is always associated with ____________ approaches.
A. Precision; nonprecision.

VOR/NDB COMPARISON MATRIX

	VOR	NDB
LINE OF SIGHT	X	
ATMOSPHERIC INTERFERENCE		X
RANGE DEPENDENT ON POWER OUTPUT	X	X
USED FOR NAVIGATION & APPROACHES	X	X
MAKES VICTOR AIRWAYS	X	

RADIO MAGNETIC INDICATOR (RMI)

Several different types of RMIs are in use by Army aircraft. The operation of the RMI as presented here is for the one used in the UH-1. Many features of the RMI, regardless of model, will remain the same.

Q. Describe the Radio Magnetic Indicator (RMI).
A. The RMI is a multi-function navigational instrument which consists of a compass card, a heading index, two bearing pointers, and a pointer function switch. It enables simultaneous determination of the present magnetic heading and the direction to or from the navigational facility or facilities which the Number 1 and/or Number 2 bearing pointers are coupled.

The compass card is moved by the aircraft's slaved gyro compass system. When working properly, the card shows the magnetic heading under the heading index. The compass card can be moved manually using the annunciator knob.

The face of the RMI has fixed reference points at 45° increments to assist in identifying turns 45° and 90° from the present heading.

A "SET HDG" knob can be used to set a heading "bug" to the desired heading.

NOTE: Always cross-check the compass card on the RMI with a known heading (mag compass) during preflight to insure proper indications are provided during flight.

Q. The Number 1 bearing pointer (the thin needle) is normally coupled to the _______ receiver, while the Number 2 bearing pointer (the fat needle) is normally coupled to the _________ receiver.
A. ADF (NDB); VOR.

Q. What is the purpose of the pointer function switch on the RMI?
A. It is used to select between the ADF and VOR receivers for aircraft equipped with more than one NAVCOM receiver. When "ADF" is selected the Number 1 bearing pointer is coupled to the ADF receiver; when "VOR" is selected the Number 2 bearing pointer is coupled to the VOR receiver.

Q. On the UH-1, what position should the pointer function switch be placed?
A. It should be left in the "ADF" position.

Q. Explain how the RMI is used to determine the following navigational information:

a. magnetic course to the navigational facility
b. radial inbound to the VOR
c. magnetic bearing outbound from the NDB
d. magnetic course inbound to the NDB
e. radial outbound from the VOR

A. a. Regardless of your present heading, the <u>magnetic course to</u> a navigational facility, either VOR or NDB, will always be the heading under the head of the appropriate needle (if you make a turn to this heading, keep in mind that without wind drift correction the magnetic course to the facility will keep changing as you are "blown" off course. In such cases, a slightly different heading may be required in order to <u>maintain</u> or "track" the original magnetic course).

b. When referring to a VOR we generally talk in terms of radials, either

inbound to or **outbound from** the facility (remember VORs radiate signals in 360 different directions like spokes from a wheel - hence the term "radials"). The radial inbound to a VOR is the same as the magnetic course to the VOR and is determined by looking under the head of the Number 2 needle. Keep in mind that since VOR compass roses on navigational charts are printed in terms of radials **outbound**, the radial inbound to the VOR will be the reciprocal of that indicated on the chart.

c. Unlike VORs, which employ a sort of common sense logic, an NDB will either be a **magnetic course to (inbound)** or a **magnetic bearing from (outbound)**. In the case of a magnetic bearing outbound from an NDB, you need only look under the tail of the Number 1 needle.

d. The magnetic course inbound to an NDB can be determined by looking under the head of the Number 1 needle.

e. The radial outbound from a VOR can be determined by looking under the tail of the Number 2 needle. This is the current radial you are on or passing through.

NOTE: If you want to know your position relative to a navaid, look under the tail of the appropriate needle.

VOR/NDB NAV MATRIX - where to look

	VOR	NDB
MAGNETIC COURSE TO	HEAD/2	HEAD/1
RADIAL INBOUND	HEAD/2	
RADIAL OUTBOUND	TAIL/2	
MAGNETIC COURSE INBOUND		HEAD/1
MAGNETIC BEARING FROM		TAIL/1

Q. Explain how to determine the location of an aircraft (cardinal direction) in relation to the tuned station.

A. Mentally divide the RMI compass card into 8 "slices" with each slice being 45° in width. A "slice" should be centered on each of the four cardinal directions (N, E, S, and W) so that there are 22½° on either side of centerline (the remaining four slices represent NE, SE, SW, and NW). Observe the tail of the appropriate needle (#2 needle for current VOR radial; #1 needle for current NDB bearing). Whichever "slice" the tail of the needle lies in represents your location in relation to the tuned station.

As an example, the tail of the #2 needle is on 290. This means that you are on the 290° radial of the tuned VOR and you are west of the tuned station (290 is within 22½° of 270 or west). If, on the other hand, the tail of the needle were on 295, then you would be on the 295° radial of the tuned VOR and you would be considered northwest of the tuned station (295 is not within 22½° of 270

<west> or 360 <north>; it is in between, thus - northwest).

Q. How is station passage indicated by the RMI?
A. By a reversal of the needle.

Q. What is considered to be the standard intercept angle when turning to intercept a course, radial, or bearing?
A. 45°

Q. If you are given instructions to "expedite" the interception of a specific course, radial, or bearing, what angle of intercept should you use?
A. 90°

Q. Although 45° and 90° are used as standard intercept angles, the aviator may use a different angle if he deems it appropriate, provided it will not conflict with the air traffic clearance issued by ATC. What four things must the aviator consider when deciding how to intercept?
A. - The angular difference between present position and new track.
- The distance from the tuned station.
- ATC clearance requirements.
- Winds.

Q. What does the acronym "OPTIC" stand for?
A. This is a sequence of events for course intercept:
(O)rientation
(P)osition
(T)racking
(I)ntersection fixing
(C)ourse interception

Q. Explain the procedure for using the RMI to determine course intercept.
A. For an NDB, regardless of direction (magnetic course to or bearing from), use the following method:

- Count 45° (standard) or 90° (expedite) from the desired course/bearing toward the head of the #1 needle. ALWAYS COUNT IN THE DIRECTION CLOSEST TO THE HEAD AND ALWAYS BEGIN THE COUNT FROM THE DESIRED COURSE/BEARING AND NOT THE ACTUAL HEADING.

For a VOR, the method to be used depends on whether you are intercepting a radial inbound or a radial outbound. If you are intercepting a radial <u>outbound</u> the method is the same as for the NDB:

- Count 45° (standard) or 90° (expedite) from the desired radial toward the head of the #2 needle. ALWAYS COUNT IN THE DIRECTION CLOSEST TO THE HEAD AND ALWAYS BEGIN THE COUNT FROM THE DESIRED RADIAL AND NOT THE ACTUAL HEADING.

If you are intercepting a radial <u>inbound</u> you must first determine the reciprocal radial, then the procedure becomes the same as the NDB or VOR-outbound:

- From the reciprocal radial, turn 45° (standard) or 90° (expedite) toward the head of the #2 needle. ALWAYS COUNT IN THE DIRECTION CLOSEST TO THE HEAD AND ALWAYS BEGIN THE COUNT FROM THE DESIRED RECIPROCAL RADIAL AND NOT THE ACTUAL HEADING.

REMEMBER: The phrase "radial inbound" in a problem should be a clue that a conversion will be necessary (reciprocal radial).

Q. What is meant by "tracking"?
A. It is a procedure of staying on a desired course by applying drift correction to the aircraft heading to counteract crosswinds.

Q. When attempting to intercept a course, radial, or bearing, or while tracking a course, radial, or bearing what is the general rule of thumb you should remember for "positioning" the respective needles?
A. "PUSH THE HEAD, PULL THE TAIL" (of the appropriate needle).

NOTE: The phrase "The head always drops and the tail always rises" is used by many aviators to remember the "positioning" for the respective needles.

Q. If you are flying inbound or outbound on a specific course, radial, or bearing and are blown off course by the wind. How can you quickly determine which direction the wind is coming from?
A. Whichever side the head of the respective needle is on is the direction the wind is blowing from.

Q. In a situation similar to the above scenario and using only information obtained from the RMI, your IP asks you what your position is relative to the desired course/track. What do you tell him/her?
A. This question may be answered in one of two ways. You can tell him/her that the course lies to your left/right; or you can say that you are left/right of course or track. Remember that the course lies to the same side as the head of the needle and the tail of the needle indicates what side of track you are on. The answers will always be opposite (ie, if the course is on your right then you are left of track, and vice-versa) so be careful about how you phrase your answer.

Q. If you intercept a desired course and notice a few minutes later that you are being blown off course by the wind, what should be your initial course re-intercept heading?
A. For airspeeds less than 90 kts, turn 30° from the desired course (your current heading) in the direction the wind is coming from (side that the head of the needle is on).
For airspeeds equal to or greater than 90 kts use 20°.

Q. You have applied the above wind drift correction heading to re-intercept the desired course, radial, or bearing. What should be your actions upon re-

intercepting the desired course?
A. If you maintain the initial re-intercept heading you will inevitably fly through the desired course. To avoid this you must apply what is known as a "follow-on" wind correction heading (also known as a "trial drift correction"). To apply the follow-on wind correction heading, the heading should be changed by turning toward the course by half the amount of initial correction.

For instance, if you used a re-intercept angle of 20°, then turn 10° back toward the course once it is re-intercepted. If you are once again blown off course you should re-intercept using the procedure in the above question, then apply a different follow-on wind correction heading which is about half again as much as before (for this example try 15° instead of 10°).

If, on the other hand, you find that the 10° trial drift correction is too much, then you need only turn parallel to the desired track and let the wind blow you back on course. After being blown back on course select a follow-on heading that is half the initial correction (or 5° in this case). Using this bracketing technique you will eventually determine the amount of wind correction necessary to maintain track in those wind conditions.

NOTE: The tracking procedures outlined here and in student handouts are recommended procedures. A wide range of wind, airspeed, and relative-position possibilities may require the use of non-standard techniques. If, for instance, your airspeed were low and the crosswind very high, then you could expect to have to use an intercept angle greater than 30°.

Q. In your cross-check you notice that the attitude indicator shows you in a slight bank yet the compass card on the RMI does not indicate a turn is being made. You glance briefly at the magnetic compass to verify that you are indeed in a slight turn. This indicates that you have what is known as a "stuck card" or "stuck compass". How would you determine what radial you were on with respect to a tuned VOR?
A. As always, to determine what radial you are on simply look under the tail of the #2 needle. Likewise, to determine the magnetic course to the VOR simply look under the head of the #2 needle. Keep in mind that in a stuck compass situation the #2 needle will continue to report accurate information with respect to the RMI compass card; however, it will not necessarily be pointing in the right direction with respect to the aircraft's actual heading. If, for instance, the tuned VOR is directly ahead of you and the magnetic compass indicates that you are heading 180, but the RMI compass card is stuck on 090 then the head of the #2 needle will appear to indicate that the tuned station is to your right, on 180, instead of directly ahead of you.

Q. How would you determine the magnetic course to a tuned NDB with a stuck compass?
A. Unlike the VOR, the head of the NDB bearing pointer (#1 needle) will point directly at the tuned station regardless of the status of the compass card. With a stuck card, the actual magnetic course to the tuned NDB can be determined by the formula: $MH +/- \Theta = MC$. Here, "MH" is the magnetic heading shown on the magnetic compass; "Θ" represents the number of degrees left or right of the heading index to the head of the #1 needle; and "MC" is the magnetic course to

the tuned NDB. If the head of the #1 needle is to the right of the heading index (right side of the compass card) then you would add Θ to the magnetic heading. If the head is on the left of the heading index (left side of the compass card) then subtract Θ from the magnetic heading.

NOTE: Some questions may ask what heading you should turn to in order to proceed directly outbound from the NBD. This is to see if you are paying attention. Since you want to determine the proper heading to fly away from the NDB then you would have to use the tail of the needle for calculations instead of the head.

COURSE DEVIATION INDICATOR (CDI)
and
VOR PROCEDURES (enroute)

VOR navigational signals are displayed on an instrument called the Course Deviation Indicator (CDI). Several different types of CDIs are in use by Army aircraft. The UH-1 uses the ID-1347, so its use and functionality will be the basis of this section. Many features of the CDI, regardless of model, will remain the same.

Q. Describe the Course Deviation Indicator (CDI)/ID-1347.
A. The CDI consists of an on-course index, a course card, a reciprocal course index, a TO-FROM indicator, an omnibearing selection knob, a course deviation needle, a glide slope indicator (ILS only), a blue/yellow off-course indicator (ILS), and two warning flags.

The course card resembles a compass card with 360° in 5° increments around its face. The course card is completely independent of aircraft heading and can be rotated with the omnibearing selector (OBS). The OBS is used to "dial-in" a desired course under the on-course index at the top of the CDI. The reciprocal course index can then be read at the bottom of the CDI.

The TO-FROM indicator is a sensing element that "senses" whether or not the "dialed-in" course will take you toward or away from a tuned station.

The course deviation needle will swing left or right (up to 10° either side) to indicate the degrees of deviation from the desired course.

In addition, there are two warning flags. One is labeled "GS" and is for the glideslope; the other is labeled "LOC" to indicate system status.

Q. At what two times can you expect to see the "LOC" warning flag?
A. (1) When the set is OFF.
(2) When the set is ON and a faulty or unreliable signal is being received.

Q. What could be the cause of an "unreliable" signal (see above answer)?
A. - The aircraft is too low for line-of-sight reception from station.
- The aircraft is too far away from the tuned station.
- Receiver malfunction.
- The VOR is undergoing maintenance.

Q. What does a horizontal bar on the face of the course indicator represent?
A. Glide slope (for ILS only).

Q. At what times can you expect to see the "GS" warning flag?
A. (1) When you ARE NOT on an actual ILS approach.
(2) When you ARE on an ILS approach and an unreliable glideslope signal is received.

Q. Is it possible for you to be flying away from a VOR and get a "TO" indication on the CDI?
A. Yes. Remember that the course "dialed-in" under the on-course index is independent of aircraft heading. If you are northeast of a tuned station and heading due north then you are obviously flying away from the station. However, if you had a heading of 245 dialed-in under the on-course index then you would

---

get a "TO" indication on the CDI. The CDI is, in essence, saying: "If you fly a heading of 245° from your present position, it will take you closer TO the station." If, on the other hand, you dialed-in the reciprocal (065°) under the on-course index, then you would get a "FROM" indication and the CDI would be saying: "If you fly a heading of 065° from your present position, it will take you farther away FROM the station.

Q. How is station passage indicated on the CDI?
A. By a reversal of the "TO-FROM" indicator.

Q. What is the purpose of the vertical needle of the course indicator?
A. The needle, which pivots at the top and swings left or right, represents the course set under the course selector index and indicates if the aircraft is flying that course. When the needle swings to the right, it means that the course is to the aviator's right; likewise, when the needle swings left, it means that the course is to the aviator's left.

Q. The center circle on the course indicator, sometimes referred to as the "doughnut", has a row of horizontal dots to the right and left. Explain the purpose of these.
A. The edge of the doughnut and the dots represent the number of degrees between the position of the aircraft and the selected course. The edge of the doughnut represents 2° off and each successive dot represents an additional 2° off.

Q. What does a full-scale deflection of the course needle indicate?
A. That you are 10° or more from the desired course.

Q. What is the exception to the above rule?
A. When performing an ILS approach, the edge of the doughnut and each successive dot represent ½° off course with a full-scale deflection indicating that you are 2½° or more off course.

NOTE: Total needle sensitivity = 20° for VOR (10° either side of centerline)
5° for ILS (2½° either side of centerline)

Q. What are the three recommended steps for using the CDI to <u>ORIENT</u> yourself with respect to a station?
A. - Tune and identify the station.
- Rotate the course selector knob until the needle is centered and the "TO-FROM" indicator shows "TO".
- Read the course TO the station under the course selector index (top) and the radial FROM the station under the reciprocal course index (bottom).

NOTE: If, for instance, you tuned the VOR and the needle centered with a heading of 090 under the course selector index and a "TO" indication, then you can say to yourself: "<u>If</u> I were flying a heading of 090°, I would be on course and I would be flying toward the station." Since you can also see from the reciprocal

---

index at the bottom that you are on the 270° radial, then you can deduce that you are west of the station (and heading in the direction indicated by your RMI or magnetic compass).

Q. What steps should be followed when attempting to FIX an intersection?
A. - Tune and identify the station to be used for the fix.
- From the ELA chart determine the radial to be used for the fix and rotate the course selector so that this value is under the course index (this will result in a "FROM" indication).
- Visualize the station you are tuning with respect to your position (ie, Is the station to the left or right of you?).
- Use the following rule of thumb: "SAME SIDE, SHORT."

In other words, if the course needle deflects to the same side as the tuned station then you will always be short. If the needle centers then you are at or over the intersection, and if the needle deflects to the opposite side of the tuned station then you are past the fix or intersection.

NOTE: The "FROM" indication is very important here. If you do not have a "FROM" indication then the opposite of the above will be true. Some test questions may try to fool you in this manner. Be careful!

REMEMBER: To ORIENT yourself use a TO indication; to FIX yourself use a FROM indication. ORIENT-TO, FIX-FROM (OT-FF)

Q. Tracking can be accomplished using the CDI in a manner similar to the RMI. However, to make tracking easier the CDI should be set up so that it is "pictorially correct." What is meant by "pictorially correct" and how can you ensure that the CDI is set up in this manner?
A. It is easier to answer this by illustrating an example. Suppose I am flying inbound to a tuned VOR on the 090 radial. My heading would be 270° and if I rotated the course selector until the needle was centered with a "TO" indication, then 270 would also be under the on-course index. If I were subsequently blown off-track by a wind from the south then the needle would deflect to the left, indicating that the desired course is to my left. Here, the CDI is "pictorially correct." However, if everything remained the same except that I had the reciprocal (090) dialed-in under the on-course index then the needle would deflect to the right. This could give you the erroneous idea that the course lies to your right because it is not "pictorially correct." The fact is, the CDI is saying: "If I were flying a heading of 090° (instead of 270°), then the desired course would be to my right."

In order to stay "pictorially correct," the general rule of thumb is to ensure that the value under the course index of the CDI is in general agreement with your actual heading. If you track inbound to a VOR and your habit is to ensure that the CDI is tuned such that you have a "TO" indication then you will have nothing to worry about.

Q. If you are flying inbound or outbound on a specific course, radial, or bearing

---

and are blown off course by the wind. How can you quickly determine from the CDI which direction the wind is coming from?
A. Assuming that the CDI is set up so that it is pictorially correct, the wind will always be coming from the side the needle is deflected to.

Q. In a situation similar to the above scenario and using only information obtained from the CDI, your IP asks you what your position is relative to the desired course/track. What do you tell him/her?
A. Again, this type of question may be answered in one of two ways. You can tell him/her that the course lies to your left/right; or you can say that you are left/right of course or track. Remember that the course always lies to the same side that the needle is deflected to. If the needle is deflected to the left then you can say that the course is to your left or that you are right of course or right of track.

Q. If you intercept a desired course and notice a few minutes later that you are being blown off course by the wind, what should be your initial course re-intercept heading?
A. For airspeeds less than 90 kts, turn 30° in the direction the wind is coming from (side that the needle is deflected to).
For airspeeds equal to or greater than 90 kts use 20°.

Q. You have applied the above wind drift correction heading to re-intercept the desired course, radial, or bearing. What should be your actions upon re-intercepting the desired course?
A. If you maintain the initial re-intercept heading you will inevitably fly through the desired course. To avoid this you must apply what is known as a "follow-on" wind correction heading (also known as a "trial drift correction"). To apply the follow-on wind correction heading, the heading should be changed by turning toward the course by half the amount of initial correction.
For instance, if you used a re-intercept angle of 20°, then turn 10° back toward the course once it is re-intercepted. If you are once again blown off course you should re-intercept using the procedure in the above question, then apply a different follow-on wind correction heading which is about half again as much as before (for this example try 15° instead of 10°).
If, on the other hand, you find that the 10° trial drift correction is too much, then you need only turn parallel to the desired track and let the wind blow you back on course. After being blown back on course select a follow-on heading that is half the initial correction (or 5° in this case). Using this bracketing technique you will eventually determine the amount of wind correction necessary to maintain track in those wind conditions.

NOTE: Notice that the intercept angles and follow-on corrections for tracking using the CDI are the same as for the RMI. Also keep in mind that these angles are the <u>recommended</u> angles and that given the wide range of wind, airspeed, and relative-position possibilities, these standard tracking procedures may require modification.

COURSE DEVIATION INDICATOR (CDI)
and
VOR PROCEDURES (enroute)

Q. Federal Air Regulations (FARs) require that VOR receivers, if used for IFR flight, receive an operational check how often?
A. Within the 30 days preceding the flight.

NOTE: A safe practice is to make an operational check of VOR equipment prior to each IFR flight.

Q. An operational check of VOR equipment should include what four things?
A. - An accuracy check of the course selector and needle.
- A check to insure proper sensing of the "TO-FROM" indicator.
- A check to insure proper reaction of the warning flag.
- A check to insure 10° of needle swing from the center to each side.

Q. What are three ways that you can test your VOR receiver/equipment?
A. - Use an authorized VOT (VOR Test Facility).
- Use an authorized VOR Airborne Check Point.
- Use an authorized VOR receiver check point on the ground.

Q. How do you determine where VOTs are available?
A. By referring to the IFR Enroute Supplement. The letters "VOT" will be followed by a frequency and can be found adjacent to "NAVAIDS" for the appropriate airport.

Q. Where can you look to find a list of VOR Airborne Check Points?
A. In the AP/1, Chpt 3, under "VOR RECEIVER CHECKPOINTS."

Q. Where would you look to find out about authorized VOR receiver check points on the ground?
A. You should check at operations for the airport or with Flight Service.

Q. Describe the procedure for checking VOR equipment using a VOT.
A. From GP, Chpt 5:
- Tune the receiver to the designated frequency of the test facility.
- Identify the facility. Two means of identification are used. In some cases, a continuous series of dots are used; while in others, a continuous 1020 MHz tone will identify the test signal.
- Ensure that the "LOC" warning flag has disappeared from view.
- With the CDI needle centered, the course selector should read 0° with a "FROM" indication or 180° with a "TO" indication (see tolerances below).
- Rotate the selector knob to ensure that the course needle will swing 10° left and right of center.

Q. What is the allowable tolerance (error) for a VOR receiver ground check?
A. +/- 4° (in the preceding question, if the course needle were centered and the course selector read anything between 356° and 004° (with a "FROM" indication), then the receiver would be within tolerance and acceptable for IFR flight).

COURSE DEVIATION INDICATOR (CDI)
and
VOR PROCEDURES (enroute)

Q. What is the allowable tolerance (error) for airborne VOR receiver checks?
A. +/- 6°.

Q. When performing a VOR receiver check, it is also a good idea to check the VOR needle (#2 needle) on the RMI. FARs do not specify a tolerance or error permissible for needle swing, but a good rule of thumb is ______° to ______°.
A. 8° to 12°.

Q. You attempt to tune-identify a VOR station (not a VOT), and all you hear are a series of continuous dots. What does this mean? Can that facility be used for navigation?
A. The facility is undergoing maintenance. When a facility is operational it is required by the FCC to radiate an identification signal. Even though it may be, in fact, operational, the signal is considered unreliable until maintenance is complete. During such periods the normal facility identification is replaced by a test signal, which consists of a series of dots transmitted for 3/4 of an identification cycle with the interval between dots equal to the time of each dot.

Q. What is meant by a "NO-NOTAM" preventive maintenance procedure?
A. This refers to routine preventive maintenance on navigational facilities.

- For locations with a single instrument approach aid, no NOTAMs are required to be posted as long as sky conditions remain scattered (no ceiling) and the visibility remains at or above 5 SM.
- For locations with two or more instrument approach aids, no NOTAM action is required as long as the ceiling remains at or above 3000' and the visibility remains at or above 5 SM.

Q. Where in DOD FLIPS will you find information regarding "NO-NOTAM PREVENTIVE MAINTENANCE PROCEDURES"?
A. In the back of the IFR Enroute Supplement (Section C).

NOTE: Equipment must be immediately returned to operation or NOTAM action taken if weather conditions deteriorate below ceiling or visibility requirements or if service directive time periods are exceeded.

HOLDING

Q. What is a holding pattern?
A. A uniform track designed to hold an aircraft temporarily in the vicinity of a fix.

Q. Give three examples of a fix.
A. VOR, NDB, intersection.

Q. There are two basic types of holding patterns. What are they and explain the difference.
A. (1) Standard - right hand turns in a racetrack pattern.
   (2) Nonstandard - left hand turns in a racetrack pattern.

Q. A holding pattern may be entered in one of three ways. What are they?
A. Direct, parallel, and teardrop entries.

Q. Using the two figures below, answer the following questions.

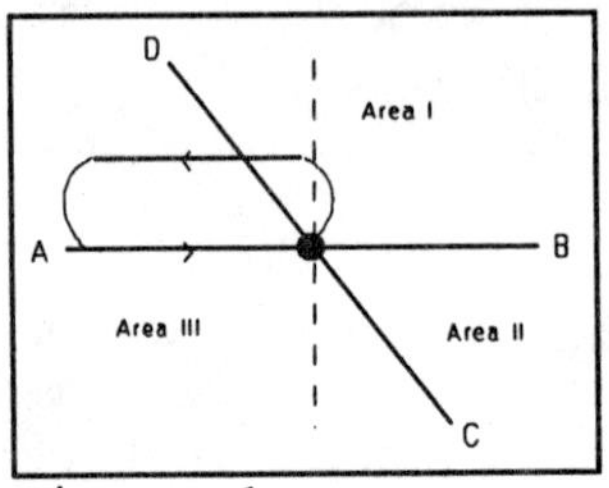

Figure 1

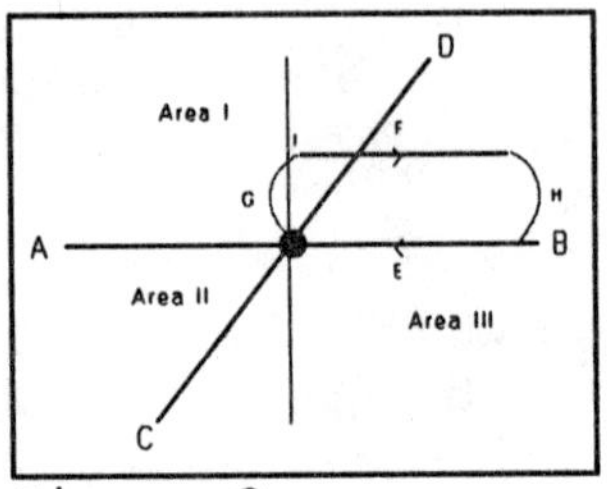

Figure 2

(1) Which figure represents a standard holding pattern?
(2) Identify the parts of the holding pattern in Figure 2 (points E - I).
(3) What should be the time inbound on leg E?
(4) What should be the time outbound on leg F?
(5) What rate of turn should be used for the turns at points G and H?
(6) If this rate of turn is maintained, how long will it take to complete the turn?
(7) At what points in the holding pattern of Figure 2 should you be interested in starting or stopping your time?
(8) Where does an inbound holding course always terminate?
(9) Line A-B in both figures represents an imaginary line that always coincides with what part of the holding pattern?
(10) Line C-D in both figures also represents an imaginary line. What is its purpose?
(11) **Area I** of Figure I represents the area bounded by B, D, and the holding fix. If your approach to holding is from anywhere in this area what entry should be used?
(12) **Area II** of Figure I represents the area bounded by B, C, and the holding fix. If your approach to holding is from anywhere in this area what entry should be used?
(13) **Area III** of Figure I represents the area bounded by C, D, and the holding fix. If your approach to holding is from anywhere in this area what entry should be used?
(14) In Figure 2, **Area I** is bounded by A, D, and the holding fix; **Area II** by A, C, and the holding fix; and **Area III** by C, D, and the holding fix. Give the appropriate type of entry for an approach from each of these areas.
A. (1) Figure 2, all turns are to the right.
   (2) E - inbound leg

F - outbound leg
G - fix-end turn
H - outbound-end turn
I - abeam point

(3) One minute.

(4) First outbound leg should be one minute; each successive outbound leg should be adjusted (for winds) to make the inbound leg one minute.

(5) Standard rate of turn - 3° per second (however, the rate of turn on the outbound-end turn, H, may require some adjustment for winds in order to properly intercept the inbound holding course -- bank angle should not exceed 30°).

(6) One minute.

(7) Upon entering holding, cross the fix and begin your turn (right or left depending on type). Once the abeam point is reached start the clock and proceed outbound for one minute. Use a standard rate turn to intercept the inbound holding course. Once wings are level, begin the inbound time. Check the time again when over the fix. If the inbound time is not one minute, adjust the outbound leg as necessary to make the next inbound trip one minute.

To sum up: outbound leg - start clock at abeam point; stop clock at one minute (first leg), or predetermined time (successive legs). Inbound leg - start clock on wings level inbound; stop clock over fix (should be one minute if outbound leg is properly adjusted). Use the time during turns to reset clock.

NOTE: It is common for flight students to get disoriented after a couple of turns in the holding pattern (particularly when holding at an intersection). "Am I going inbound or outbound? What was my heading?" Of course, the TO-FROM indication will help when holding over a navaid, but what about an intersection? It is a useful technique, for both you and your copilot, to say out loud what you are doing. For instance, "I have positive station passage, right turn to 090° outbound." Upon reaching a heading of 090°, "Wings level heading 090°, start the clock, looking for 55 seconds." And, after proceeding outbound for 55 seconds, "55 seconds, right turn to 270° inbound, looking for one minute." Etc, etc. Constant communication reinforces what you are doing and keeps your copilot informed (this is known as Crew Coordination).

(8) At the holding fix.

(9) The inbound holding course - remember that the inbound holding course will be the reciprocal of the radial you are holding on if over a VOR.

(10) You will reach a point where you will recognize the appropriate entry for holding without the use of aids. There are several techniques (only one of which will be covered here). Find the one that works best for you and use it.
Many students find it useful to use an overlay, particularly on tests, to determine the appropriate entry. As mentioned earlier, the three types of entry are direct, parallel, and teardrop. The type of entry used depends on your position relative to the holding pattern.

Step 1: Look at, or draw the appropriate holding pattern keeping in mind that the inbound leg always terminates at the holding fix. Make sure all turns are right for a standard holding pattern and left for a nonstandard holding pattern.

Step 2: Draw or visualize the line A-B that coincides with the inbound leg.

Step 3: Draw or visualize the line C-D. The angle between line C-D and A-B should be 70° (or you can visualize the line as being 30° offset from a vertical line passing through the fix (dashed line above)). If using the 30° offset from vertical, go to the right for right turns (standard); go to the left for left turns (nonstandard).

Step 4: Notice from the above diagrams that the entire pattern can now be divided into three sectors or areas.

The side of line C-D that contains the holding pattern is the **largest area** (Area III). An approach from anywhere in this area should be a **direct entry.**

The side of line C-D that does not contain the holding pattern is further divided into two areas, the **smallest** (Area II) always being a **teardrop entry.** The only **remaining area** (Area I) represents a **parallel entry.**

In the examples presented above, the vertical line passing through the fix is oriented straight up (due north on a map). Don't expect this to be the case every time as every holding pattern will be different. Follow the steps above and you will not go wrong.

(11) Parallel
(12) Teardrop
(13) Direct
(14) Area I - Parallel
Area II - Teardrop
Area III - Direct

Q. The side of the inbound course on which the holding pattern lies is known as the maneuvering side or the holding side. For a standard holding pattern, the holding side is always to the ___________ (left/right) of the inbound course.
A. Right.

Q. For a nonstandard holding pattern, the holding side is always to the _________ (left/right) of the inbound course.
A. Left.

Q. If you were holding on an inbound course of 360° with a direct headwind of 10 knots, how would you expect this to effect your outbound leg?
A. The outbound leg should be something less than one minute in order to make the inbound leg exactly one minute.

Q. There are two types of holding clearances. What are they?
A. Detailed and abbreviated.

Q. What is an example of an abbreviated holding clearance?
A. "Hold as published...."

Q. What six items of information should be included in a detailed holding clearance?
A. (1) **ID** (ie, "R-12345....")
(2) **Direction** from holding fix (ie, "...hold west of....")

(3) **Name** of holding fix (ie, "...the Macon VOR...")
(4) Holding **radial, bearing or airway** (ie, "...on the 279° radial...")
(5) **Type** holding pattern (this will be given **ONLY** if the pattern is nonstandard) (ie, "...left turns....")
(6) **Expect Further Clearance (EFC)** time (ie, "...time now 25, expect further clearance at 35.")

Q. You are flying direct to the Macon VOR. You have tune-identified the station and the head of the #2 needle reads 060°. Using the detailed holding clearance from the previous question, what type of entry would you use?
A. Direct; using the 4 steps outlined above:

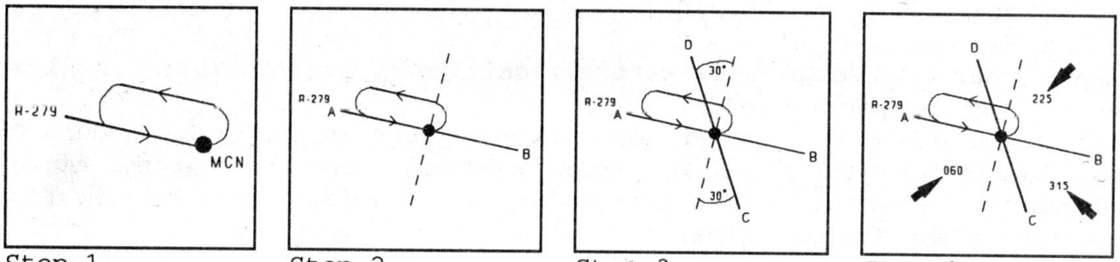

Step 1 Step 2 Step 3 Step 4

Step 1: Draw the holding pattern. Notice that the body of the holding pattern lies "west" of the MACON VOR on the 279° radial. Also notice that the inbound leg terminates at the VOR and that all turns are to the left. Keep in mind that your inbound heading for holding will be the reciprocal of 279°, or 099°.

Step 2: Draw the line A-B so that it coincides with the inbound heading. Also draw a dashed line perpendicular to line A-B and running through the fix.

Step 3: Draw the line C-D. Since turns are to the **left**, move 30° **left** from the dashed line drawn through the fix.

Step 4: Your heading is 060°. The tail of the #2 needle confirms that you are SW of the VOR. This puts you on the side of line C-D that contains the holding pattern. This means that your entry will be direct.

If, on the other hand, your heading to the VOR were 315°, the tail of the #2 needle would confirm that you are SE of the station. This places you in the small region bounded by B, C, and the holding fix. In this case, the correct entry would be a teardrop.

One final example, your heading is 225° direct to the VOR. Here, the tail of the #2 needle would confirm that you are NE of the station. This would place you in the area that is neither direct nor teardrop, but parallel. See, this really is easier than it first looks!

NOTE: Some holding patterns are published on IFR charts. If ATC requests that you hold at such an intersection, you will be expected to hold as published unless otherwise advised by ATC. You can also expect to receive an abbreviated clearance or one that is less detailed.

Q. For what reasons might ATC require you to hold at a particular fix?
A. You may receive a clearance limit, a fix beyond which you may not proceed without further clearance from the appropriate agency. If you reach the clearance limit without receiving further clearance then you must hold until such clearance is received. Holding is one way that ATC can easily control the flow of traffic into or out of heavily congested areas or to "hold" you in one controller's airspace while coordination is made for clearance into an adjacent controller's airspace.

Q. If you reach your clearance limit before receiving clearance beyond the fix what is the recommended procedure?
A. If the holding pattern is charted, maintain the last assigned altitude and begin holding IAW the pattern depicted. If no holding pattern is charted, begin holding a standard holding pattern on the course on which you approached the fix and immediately request further clearance.

Q. Holding instructions from ATC that contain only the holding direction should be an indication of what?
A. The holding pattern is charted - refer to the appropriate charts.

Q. If an ATC controller desires you to hold, how far in advance of the fix should you receive holding instructions?
A. At least 5 minutes prior.

NOTE: If ATC feels it will be necessary for you to hold short of a clearance limit, the holding instructions will normally be included in the takeoff clearance.

Q. If you are within 5 minutes of your clearance limit or designated holding fix and have not received holding instructions, what should you do?
A. Request holding instructions from ATC.

Q. If you are holding as instructed and the expect further clearance time comes and goes without you hearing from ATC, what should you do?
A. Contact ATC and request further clearance.

Q. What reports are required upon reaching an assigned holding point or clearance limit?
A. The time of arrival and altitude to the holding fix or point to which cleared (ie, "Indy Center, this is Army 12345, Chesty, 42, 5000").

NOTE 1: You must also report leaving an assigned holding point to ATC.

NOTE 2: The reporting requirement for pilots reaching or leaving an assigned holding fix may be omitted by pilots of aircraft involved in instrument training at military terminal area facilities when radar service is being provided.

Q. Once you receive clearance to depart a holding pattern, at what point in the pattern should you actually depart?
A. After crossing the fix on the inbound leg.

Q. What is the exception to the above answer?
A. You receive ATC clearance for a "present position" departure (ie, "...you are cleared present position direct to...").

Q. A direct entry into holding is fairly self-explanatory. Describe the procedure for parallel and teardrop entries.
A. For a parallel entry, turn the shortest direction to the outbound heading and parallel the outbound course. After flying outbound for one minute, turn back to intercept the inbound course.

For a teardrop entry, cross the holding fix and turn to a heading within a 30° track from the inbound holding course (inside the holding pattern). Fly this heading for one minute and then turn to intercept the inbound holding course.

Q. If holding at a navaid becomes necessary, ATC will almost always clear you to hold on your current inbound heading. In this case, your entry will always be ____________.
A. Direct.

Q. As mentioned earlier, time for the outbound leg should begin when you are abeam the holding fix. What is perhaps the quickest way to determine when you are abeam the fix? What is the most accurate way?
A. In the case of a navaid, when the TO-FROM indicator changes or when the appropriate needle on the RMI points to the radial 90° from the holding fix radial (check 90° reference point on the face of the RMI).

The most accurate way is to rotate the course selector knob on the CDI 90° from the inbound course in the direction of turn. When the course deviation needle centers in the doughnut, you are abeam the fix.

Q. Since it is normally not possible to fix the abeam point when holding at an intersection, when does the outbound time start?
A. When you complete the turn (wings level).

Q. After entering holding you proceed outbound for one minute on the first outbound leg. The time on the inbound leg is 40 seconds. What should be the time on your next outbound leg?
A. 1 minute and 30 seconds;

This can easily be calculated using the E6-B in one of several ways. The following is the method I use:

You know the inbound time and want to find the outbound time (necessary to make the inbound time 60 seconds or 1 minute). Therefore, use the inner scale for the inbound leg time and the outer scale for the outbound leg time.

Place "40" from the inner scale under "60" on the outer scale. Next, look at the "hour index" (represented by a large black arrow and the number "60" on the inner scale). The number above the "hour index" on the outer scale is your new time outbound - 90 seconds (or 1 minute and 30 seconds).

Q. Later, you enter holding again and proceed outbound for one minute on the first outbound leg. The time on the inbound leg is 1 minute and 55 seconds. What should be the time on your next outbound leg?
A. 31-32 seconds;
Since 1 minute and 55 seconds is 115 seconds, place "115" on the inner scale under "60" on the outer scale. Look at the "hour index". The number above the "hour index" on the outer scale is you new time outbound - between "31" and "32" seconds.

Q. Adjusting the time on the outbound leg will correct for headwinds and tailwinds. How do you correct for crosswinds?
A. By applying holding pattern drift correction (crabbing).

Q. What is the general rule of thumb for holding pattern drift correction?
A. If your correction on the inbound leg is 10° or less, use double the inbound correction (in opposite direction) on the outbound leg.
If it is more than 10°, use the inbound correction plus 10°.

NOTE: The holding pattern drift correction mentioned above is from FM 1-240. The Airman's Information Manual (AIM) recommends tripling the inbound drift correction on the outbound leg in order to avoid major turning adjustments.

Q. What is the maximum indicated airspeed allowed for holding?
A. 175 kts for all propeller-driven aircraft and helicopters (unless using a holding pattern depicted on a COPTER approach, in which case 90 kias is the maximum).

Q. When entering holding, the holding fix should be crossed, initially, at or below the maximum holding airspeed. If a speed reduction is necessary, when should it be accomplished?
A. When 3 minutes or less from the holding fix.

INSTRUMENT APPROACH PROCEDURES

Q. What are the four basic segments of an instrument approach procedure?
A. - Initial approach segment
- Intermediate approach segment
- Final approach segment
- Missed approach segment

Q. What is the purpose of a feeder route?
A. To transition aircraft from the enroute structure to the initial approach fix.

Q. What information is provided with a feeder route on the approach chart?
A. Minimum altitude, course direction, and distance.

Q. What is meant by the term "initial approach"?
A. The aircraft has departed the enroute phase of flight, and is maneuvering to enter an intermediate or final segment of the instrument approach. This includes the route of flight from the Initial Approach Fix (IAF) up to and including the procedure turn inbound (if used).

Q. What is included in the intermediate approach segment?
A. The inbound course to Final Approach Fix (FAF).

Q. Where does the final approach segment begin and end?
A. Begins at the FAF and ends at the Missed Approach Point (MAP).

Q. Where does the missed approach segment begin and end?
A. Begins at the MAP and ends at the fix for holding or return to the enroute structure.

Q. What Flight Information Publication is normally used by Army aviators for planning and executing instrument approach procedures?
A. The DOD FLIP Terminal Low Altitude U.S. Instrument Approach Procedures (IAP) booklet(s).

Q. The above DOD FLIPs are normally known by what abbreviated name?
A. "Approach plates"

NOTE: Some airfields, for instance those shown in green on an ELA chart, have approved instrument approach procedures that are not published in DOD FLIPs. These approach procedures may be found in a U.S. Gov't Terminal Procedures FLIP. The U.S. Gov't Terminal Procedures FLIP has an open binding with pre-punched holes in the top. The pages are loose-leaf but generally come with a flexible metal binding strap to keep the pages together. The publication contains instrument approach procedures, airport diagrams, SIDs, and STARs. There are two black bands near the top of each volume and should not be confused with the DOD Terminal FLIPs which are published in two volumes (East/West), have a closed binding at the top and a black band at the top and bottom of each volume.

Q. How many approach plates are there and how do you know which one to use?
A. Approach plates are published in 12 basic volumes for the United States. The

back cover of each volume has a pictorial representation of the region covered (the contiguous US for US Approach Plates). This region is further subdivided into areas showing which volumes cover which areas. In the case of US Approach Plates, each volume contains information for one or more states. The area covered by the each volume is also represented on the front cover.

Q. How often are approach plates issued?
A. Every eight weeks.

Q. What publication is published at the four-week midpoint (see above question) and contains revisions, additions, and deletions to the last complete issue of the basic volumes?
A. A Terminal Change Notice (TCN).

NOTE: When planning an IFR flight - always check publication dates on all materials you use to ensure that they are not out-of-date. When you get to the portion of the flight requiring the use of approach plates check the front cover. You will see three dates, labeled "EFFECTIVE", "TO", and "TCN EFFECTIVE". If the date is after the "TO" date then the publication is no longer current. If the volume is current but the date is later than the "TCN EFFECTIVE" date, look for the TCN that should accompany the volume. Changes may or may not be contained in the TCN for your particular approach. Find out!

Q. In addition to a TCN, if required, what else should you look for prior to making a decision about a particular approach to be flown?
A. In addition to asking yourself whether or not you have the required equipment, weather minimums, etc, for the approach to be flown, you should consult any Urgent Change Notices (UCN), which are not scheduled publications and are issued as required. Also, check the IFR Supplement for that airfield and, of course, always consult NOTAMS.

Q. The effective date of the approach procedures contained in the basic volumes and those issued in TCNs is noted on the front cover. What is the effective **TIME** of these procedures?
A. 0901Z on the date shown on the front cover.

Q. If you receive an Urgent Change Notice (UCN) or a Terminal Change Notice (TCN) what should you do with it?
A. Staple the new or revised chart to the inside front cover of the appropriate volume and make any necessary changes on the affected chart(s).

Q. What is the easiest way to tell if a TCN has changes for your intended airport of landing or departure?
A. Check to see if the airport is listed in the Table of Contents of the TCN. If it is not listed then the airport information in the basic volume has not changed.

Q. Approach charts within each volume are arranged in alphabetical order. For what reason, then, might you use the Index of Instrument Approach Procedures in

the front of the volume?
A. The Index of Instrument Approach Procedures provides a quick reference of all approaches available at a particular airport. Also, in some cases an airport may be listed under a different name. If, for instance, you were planning a flight to Hesler-Noble Field in Laurel, Mississippi, and you flipped through volume 12 to the "H"s you would not find it. However, by using the Index you would be instructed to "SEE LAUREL, MS". Looking under "LAUREL, MS" you would find that Hesler-Noble Field has both an NBD and a VOR/DME approach, as well as "TAKE-OFF MINIMUMS". To make things even simpler, the page numbers are given.

Q. All approach plates or charts follow a specific chart format. Each chart consists of five sections. What are they?
A. Margin Identification, Plan View, Profile View, Landing Minima Section (with notes), and Airport Sketch.

Q. Name some items of information that would be contained in the Margin Identification section of an approach chart.
A. - Airport name and identifier (ie, LAWSON AAF (LSF)).
- Location/Community served (ie, Ft. Benning (Columbus), Georgia).
- Reference number and approving authority (ie, AL-146 (US Army)).
- Page number
- Coordinates of airport (Lat/Long)
- Procedure identification (ie, VOR RWY 14).
- Amendment, if necessary (ie, Amdt 7).

Q. What determines the procedure identification for a particular approach?
A. This is derived from the type facility providing final approach course guidance (ie, VOR, NDB, ILS, LOC, TACAN, GPS, MLS, RNAV, SDF, LDA).

Q. When is the runway number also provided as part of the procedure identification?
A. When the approach course is within 30° of the runway centerline.

Q. In some cases an airport may have more than one approach to the same runway using the same final approach guidance. How should these procedures be identified?
A. They should be identified numerically (ie, VOR 1 Rwy 18, VOR 2 Rwy 18).

Q. Occasionally you will see a procedure identification with an alphabetical suffix (ie, VOR-A). What should this immediately tell you about the approach?
A. The procedure does not meet the criteria for straight-in landing minimums (circling minimums will always apply).

Q. If an airport has a VOR-A does this mean that there is a VOR-B?
A. No, the first such procedure must have the suffix "A" even though there is not another such procedure or plans to formulate one.

Q. What are the criteria for straight-in landing minimums?
A. - The final approach course of the instrument approach procedure must be

within 30° of the runway alignment.
- A normal rate of descent must be possible from the IFR altitude shown in the approach to the runway surface.

Q. What does a slash (/) in the procedure identification mean (ie, VOR/DME Rwy 6)?
A. The word "and" can be substituted for the slash. In this example, two types of equipment must be used to execute the approach - VOR **and** DME.

Q. If a procedure identification has the word "or" (ie, "ILS or NBD") what does this indicate?
A. Either type of equipment may be used to execute the final approach.

NOTE: Look carefully at the entire procedure before determining if the approach can be flown. You may have the necessary equipment to fly the final approach but not the missed approach (ie, DME may be required). Don't get caught in this snare!

Q. How is a procedure that is restricted to "helicopter use only" depicted?
A. With the prefix "COPTER" (ie, COPTER NDB 270). Most, but not all, COPTER procedures also have the words "COPTER ONLY" imprinted somewhere on the face of the chart.

Q. All "COPTER" procedures are restricted to what airspeed?
A. 90 kias.

Q. If "COPTER" procedures are limited to 90 kts, why do such procedures have a FAF to MAP time available for speeds greater than 90 kts (ie, 105 kts)?
A. The time required to traverse the distance from FAF to MAP is based on ground speed not indicated airspeed. With a tailwind, an indicated airspeed of 90 kts may give you a ground speed significantly higher than that indicated.

Q. You see what looks like a DOD Instrument Approach Procedures booklet but it has black hash marks across the top and around the edges. What is it?
A. It contains approaches that are for **High Altitude** use only.

Q. Some charts will have the words "**CAT II**" in parentheses at the top and bottom in the margin information. What does this mean?
A. You cannot make this approach unless you meet the specific qualifications and equipment requirements outlined in AR 95-1, Table 5-3. Most, if not all, helicopters in the U.S. Army **do not** meet this criteria.

Q. What information may be found in the Plan View portion of the approach chart?
A. Navigation and communication frequencies, enroute and feeder facilities, as well as all necessary information pertaining to the initial approach segment, including procedure turn and Minimum Safe Altitude.

Q. Normally, all information within the Plan View is shown to scale. Sometimes, information necessary for the approach must be shown but cannot be plotted to

scale. How is this accomplished?
A. The Plan View may contain from one to three concentric circles. The inner circle or ring is denoted by a solid line (normally 10-nm radius) and represents the boundary of the procedure that is charted to scale.

The next concentric circle, if plotted, is a dashed line known as the Feeder Facilities Ring. This ring provides for radio aids to navigation, fixes, and intersections that **are not part of** the enroute structure but are routinely used by ATC in the directing of traffic between the enroute structure and the initial approach fix. All information on the Feeder Facilities Ring, while not plotted to scale, are shown in their relative position.

The third concentric circle, known as the Enroute Facilities Ring, is a dashed line that provides all radio aids to navigation, fixes, and intersections that **are part of** the Enroute Low Altitude Airway structure and used in the approach procedure. The facilities/fixes will be appear on the ELA chart and are shown in their relative positions on the ring.

Q. A smaller ring, normally located in one corner of the Plan View, gives a Minimum Safe Altitude (MSA), by sector, that provides at least _______ feet obstacle clearance within ________ nautical miles of the navaid providing final approach course guidance, but does not necessarily assume acceptable reception of the navigational facility depicted.
A. 1,000; 25.

Q. The Profile View of the approach chart shows a side view of the procedure to be flown. What information would you expect to find here?
A. Minimum altitude and maximum distance for the procedure turn, altitudes over prescribed fixes, distances between fixes and the missed approach procedure. Other information such as Final Approach Fix (FAF), Glide Slope angle, and Threshold Crossing Height (TCH) may also be provided.

Q. What does "TERP" stand for?
A. TERminal instrument Procedure.

Q. TERPs provide for differences in aircraft performance by placing aircraft in one of several categories and then establishing approach minimums for each category. The Landing Minima Section contains these landing minimums. Generally speaking, the faster the approach is flown the lower the landing weather minima. How many aircraft approach categories are there and what airspeeds are associated with each?
A. Six;

APPROACH CATEGORY	SPEED (KIAS)
A	Speed less than 91 kts
B	Speed 91 kts or more but less than 121 kts
C	Speed 121 kts or more but less than 141 kts
D	Speed 141 kts or more but less than 166 kts
E	Speed 166 kts or more
COPTER	Restricted to helicopters only; 90 kts max

NOTE: The above approach categories should not be confused with any of the following:
- Approach categories based on aircraft equipment (CAT I, CAT II, etc).
- Aircraft weather categories (Cat I, Cat II, etc, which are based on aircraft weight).
- Clear Air Turbulence (CAT).

Q. What are the airspeeds in the above approach categories based on?
A. Fixed-wing aircraft have a recommended final approach speed, known as V_{REF}, which is normally 1.3 times the aircraft's stall speed in the landing configuration (V_{SO}) at maximum gross weight. Fixed-wing pilot's use their aircraft's V_{REF} as a guide for which approach category they fall under and, in most cases, do not have the luxury of choosing a lower category. On the other hand, rotary-wing aircraft may choose a slow approach (Category A) or a faster approach up to operating limits.

NOTE: An aircraft with a V_{REF} of 100 kts would fall under approach Category B. However, if the aircraft found it necessary to make the approach at 121 kts or more but less than 141 kts, approach Category C minimums would have to be used.

Q. The LANDING MINIMA SECTION has a space immediately below it for including notes and is sometimes referred to as "MINIMA DATA". This section should always be checked as it provides information necessary for the safe and successful execution of the approach. What is meant by the following symbols in the notes section:
1. Negative symbology "A" enclosed in a triangle.
2. Same symbol as above but followed by the letters "NA".
3. Negative symbology "T" enclosed in an inverted triangle.

A. (1) Alternate minimums not standard, refer to appropriate regulations.
(2) This facility may not be used as an alternate due to unmonitored facility or absence of weather reporting service (only exception is if descent from minimum altitude for IFR operations, approach, and landing can be made in VMC conditions).
(3) Refer to tabulated data for specific IFR take-off minimums and/or departure procedures for this facility.

Q. Answer #3 in the above question refers to "tabulated data". Where can this be found?
A. In the front of an approach volume immediately following the legend information.

Q. A facility in the tabulated data gives the following information: "Rwy 5, 800-6, or std with min climb of 360/NM to 1700". Explain what is meant by this.
A. AR 95-1 gives Army pilots their IFR take-off minimums. With less than 50 hours of actual weather time logged you are normally restricted to minimums of 100 - ¼ (that is, 100' ceiling and ¼-mile visibility). An aviator with more than 50 hours of actual weather time (certain restrictions apply) has no minimums. Again, these are **standard** IFR take-off minimums. At the facility in this example the take-off minimums for departing runway 5 are **non-standard** and any pilot, regardless of weather time, would be limited to an 800' ceiling and 6 miles

visibility - unless the departure could be made with a climb rate of at least 360 feet per nautical mile (360/NM) to 1700' msl - in which case the standard IFR take-off minimums would again apply.

NOTE: The purpose of this is to assist pilots in obstacle avoidance.

Q. Rates of climb in most aircraft are measured in terms of feet per minute on the VSI. What method(s) do we have for determining a rate of climb of, say, 360 feet per nautical mile (360/NM)?
A. Method 1: Inside the back cover of each instrument approach volume is a "RATE OF CLIMB/DESCENT TABLE". Find the appropriate **climb/descent gradient** along the left-hand side of the chart and move across to the appropriate **ground speed.** Notice that the climb/descent gradients are given in 50' increments. A gradient of 350/NM and a ground speed of 90 knots would be a rate of climb of 525 fpm. Since you wish to find the rate of climb for 360/NM you will have to interpolate between 350/NM and 400/NM.

Method 2: Using the circular slide rule (E6B), consider that your desired ground speed and the desired rate of climb in fpm are located on the outer scale. If you plan on a ground speed of 90 kts, place the black arrow or "hour index" on the inner scale under the "90" on the outer scale. Next, read around the inner scale until you reach the specified ft/NM (drop the last digit). In this case, read around to "36" on the inner scale. Above this number on the outer scale is "54". This indicates a rate of climb of at least 540 fpm on the VSI.

To recap, with the black arrow on the desired ground speed, read ft/NM on the inner scale and fpm on the outer scale (drop the last digit).

Q. One of the most important parts of the AIRPORT SKETCH or AIRPORT DATA section is the time and speed table from the FAF to the MAP. When should this table used?
A. Use for all non-precision approaches to determine the point at which a missed approach must be executed (if runway not in sight). Always start a clock at the FAF inbound. This is the point in the profile view designated by the "maltese".

NOTE: Whether the approach is a precision or non-precision approach, it is a good technique to always start your clock over the FAF. Take, for example, the following scenario: You are shooting an ILS approach and for some reason lose the glide slope. In this case the approach would revert to a localizer approach which is a non-precision approach. If you were not in the habit of always starting your clock at the FAF inbound, then you would have no way of knowing where the missed approach point is (assuming, of course, that you are in the clouds). In this case you would have to immediately initiate the missed approach procedure and try again - perhaps wasting valuable time and fuel.

Q. When would the time and speed table (see above question) not be included on the approach chart?
A. - When the procedure is a VOR/DME approach.
- When the facility used for final approach guidance is on the airport and the facility is the MAP (terminal approach).

Q. The time and speed table is incremented at 60, 90, 120, 150, and 180 kts ground speed. A conversion will be necessary if your ground speed does not match one of these. Explain how the E6-B can be used to make this conversion.
A. Let's say that you determine from the doppler, GPS, or the wind-face portion of the E6-B that your ground speed for the approach is 115 kts.

We are normally accustomed to placing the black arrow or "hour index" on the ground speed. However, the distance from the FAF to the MAP is normally quite small and if it is smaller than 10% of the ground speed (good rule of thumb) then the small black arrow or "seconds index" at "36" on the inner scale should be used.

So, place the "seconds index" on the inner scale to "115" on the outer scale. Next, look at the time and speed table and read the distance from the FAF to the MAP. If the distance were 9.3 NM, read around the outer scale until you get to "93" (which represents 9.3). Below this number on the inner "time" scale you will see that the time from FAF to MAP will now be 4 minutes and 50 seconds.

Q. If an airfield is equipped with runway lighting, this information will also be available in the AIRPORT DATA section. The type of lighting system will be indicated by an alphabetical or alphanumeric symbol. Where can you look to find out what the symbols mean?
A. In the Flight Information Handbook, Section B, "Approach Lighting Systems"; in the GP, Chapter 2, under "Airport Lighting"; or in the IFR Supplement under the airfield name.

Q. Perhaps the most common approach/runway lighting systems use "A", "P", and "V" as the letter identifier. What do each of these stand for?
A. (A) - Approach Lighting System (ALSF, SALS, SSALR, MALS)
(P) - PAPI: Precision Approach Path Indicator
(V) - VASI: Visual Approach Slope Indicator

Q. When the approach/runway lighting system symbol is displayed with negative symbology what does this mean?
A. This indicates that the lighting is pilot-controlled.

Q. If an approach lighting symbol is portrayed with a dot "·" above it, what does this mean?
A. Sequenced flashing lights are installed with the approach lighting system.

Q. How does a pilot control pilot-controlled lighting?
A. Communications data is contained in the upper left corner of the Plan View section of each chart. If pilot-controlled lighting is available you will see a negative symbology "L". Tune in the frequency next to this symbol and key the mike 3 to 7 times in five seconds (depending on desired illumination). Lights will normally remain illuminated for 15 minutes from the most recent time of activation.

Q. Does the VASI glide slope coincide with an ILS or PAR glide slope?
A. Normally, yes, when the ILS and PAR facilities are located on the same runway.

In cases where the glideslope angles differ, deviations will be shown in the enroute supplement under "Svc". If you see the symbol V_3 in a circle on a runway diagram then the VASI glide slope is not coincidental with the PAR and/or ILS.

Q. What indications will you receive from a VASI system during an approach?
A. - On glide path: red lights over white lights (far lights red; near white)
- Above glide path: all white lights
- Below glide path: all red lights

MEMORY AID: "RED over WHITE, you're alright; RED over RED, you're dead!"

Q. A VASI is normally aligned to provide a visual glideslope angle of _____° to _____° from the horizon.
A. 2.5; 3.0

NOTE: When conducting a VASI approach, align the aircraft with the runway or runway lights at an altitude of approximately 250-300 ft for each mile of distance from the airport (ie, at 3 miles remain at or above 750' agl until "intercepting" the glide path).

Q. What is considered to be the useful range of VASI?
A. Depending on visibility conditions, approximately ½ to 1 mile during daylight and 1 to 5 miles during darkness.

Q. What is indicated by a black star in the plan view section of an approach chart?
A. This indicates that the control tower or ATIS operates non-continuously, or that there is non-standard pilot-controlled lighting.

Q. What would a "#" symbol adjacent to a tower name indicate?
A. That the control tower is closed until further notice.

Q. A procedure turn is a maneuver prescribed when it is necessary to reverse direction to establish an aircraft inbound on an intermediate or final approach course. What are some instances when the procedure turn should not be used?
A. - When the symbol NoPT is shown;
- when RADAR VECTORING is provided;
- when a holding pattern is published in lieu of procedure turn; and
- when the procedure turn is not authorized.

Q. A procedure turn is often depicted with a "barb" to one side. What is the purpose of this?
A. The barb indicates the direction or side of the outbound course on which the procedure turn should be made.

Q. The procedure turn barb (see above question) contains headings for a 45° procedure turn. Is the procedure turn required to be executed in this manner?
A. No. The point at which the turn may be commenced and the type and rate of turn is left to the discretion of the pilot.

Q. What is indicated by the absence of a procedure turn barb in the Plan View?
A. A procedure turn is not authorized for that procedure.

Q. Discuss the procedure turn options.
A. - **45° procedure turn:** while flying the outbound course, and at the procedure turn point, turn 45° toward the procedure turn side and fly outbound for 40 seconds, after which a turn in the opposite direction should be made to intercept the inbound course. The 40-second outbound time should be adjusted if known crosswinds exist. An allowance of 1° for each degree of wind drift correction used on the outbound leg should be applied to the 40 seconds flown on the leg of the procedure turn (ie, 10° wind drift correction on outbound leg = 50 seconds flown on procedure turn leg if turn is upwind or 30 seconds if turn is downwind).

- **Tear-drop procedure turn:** a teardrop turn may be executed in lieu of the 45° procedure turn if the aircraft heading at the time of crossing the approach facility is conveniently aligned with a teardrop course. Upon arrival over the approach fix, turn 30° away from the outbound course on the procedure turn side and proceed outbound for 1 minute. After 1 minute a turn in the opposite direction should be made to intercept the inbound course. If additional time is needed to lose altitude, the time outbound may be adjusted as long as the teardrop track is also adjusted. Use the following rules of thumb:

1 minute outbound = 30° teardrop track
2 minutes outbound = 20° teardrop track
3 minutes outbound = 10° teardrop track

- **Racetrack pattern:** this procedure must be used when a **"holding pattern in lieu of procedure turn"** is published. Upon crossing the approach facility execute the procedure turn the same way you would enter holding.

- **80°-260° course reversal:** this procedure cannot be found in FM 1-240 but is described in the Airman's Information Manual (AIM). Proceed outbound on the outbound course until reaching the procedure turn point, then turn 80° from the outbound course on the procedure turn side using a standard rate of turn. Immediately upon reaching a heading 80° from the outbound course begin a turn in the opposite direction. A 260° standard rate turn in this direction should be adequate to intercept the inbound course.

NOTE: If a procedure turn is depicted as a procedure track, then the turn must be flown exactly as depicted.

Q. What rate of turn should be used for procedure turns?
A. A standard rate of 3° per second should be used but may be increased or decreased to allow the aircraft to roll out on the desired track. At no time should the angle of bank exceed 30°. For aircraft equipped with an integrated flight system which uses a steering pointer, the turn should be executed with a centered steering pointer (equivalent to approximately a 25° bank angle).

Q. What limits, if any, are imposed on procedure turn distances?
A. Procedure turn limiting distances are normally set at 10 NM for standard approaches or 5 NM for COPTER approaches. Deviations from normal will be clearly depicted on the approach charts.

When flying outbound on the procedure turn you should expect to fly a minimum of one minute before turning inbound. The outbound leg may be extended, if necessary, to lose additional altitude or to compensate for adverse wind effects, but in no case should the leg be extended beyond the limit published on the approach chart.

Q. What are the two types of instrument approaches?
A. Precision and nonprecision.

Q. What are the two types of precision approaches?
A. (1) ILS - instrument landing system
(2) PAR - precision approach radar (a ground controlled approach).

NOTE: The Microwave Landing System (MLS) is also considered a precision approach but is not used by most Army helicopters.

Q. What are the five <u>normal</u> ground components of an instrument landing system?
A. - localizer
- glide slope
- outer marker
- inner marker
- approach lights

Q. The above components of the ILS can be divided functionally into three parts. What are these three parts and what information is provided by each?
A. (1) Localizer and glide slope - provides guidance information (course and altitude).
(2) Marker beacons and/or DME - provides range information (accurate radio fix, distance).
(3) Approach lights, VASI, and runway lights - provide visual information.

Q. What may be substituted for the outer or middle markers on an ILS approach?
A. Outer marker - compass locator
precision radar
surveillance radar
DME at glide slope site (when published on ILS procedure)
Inner marker - compass locator
precision radar

Q. What is a marker beacon?
A. A radio facility capable of transmitting a signal in a vertical direction only. The signal will be received only when flying directly over the facility. Its purpose is to provide a definite radio position fix, such as on an ILS approach. ILS marker beacons have a rated power output of 3 watts or less and all marker beacons transmit on a frequency of 75 MHz. Thus, the receiver, which is preset to this frequency, will respond with the appropriate audio and/or visual indications when flying directly over the beacon.

Q. A marker beacon may be used as an outer marker, middle marker, or inner marker. How can you tell the difference between them?

INSTRUMENT APPROACH PROCEDURES

A. The outer marker may be identified with continuous dashes; the middle marker by alternating dots and dashes; and the inner marker by continuous dots.

OUTER MARKER (OM)	- - - - -
MIDDLE MARKER (MM)	- · - · -
INNER MARKER (IM)	· · · · ·

Q. What is a compass locator?
A. They are normally L-class non-directional beacons (used as a component of the ILS system) having a power output of less than 25 watts and reliable to approximately 15 NM. All L-class beacons have a two-letter identifier.

Q. What is a locator marker beacon?
A. A compass locator co-located with a marker beacon. Compass locators are often situated at the outer marker site and occasionally at middle marker sites.

Q. The compass locator portion of the locator outer marker will always transmit what identification?
A. The first and second letters of the localizer ID.

NOTE: The compass locator portion of the middle marker, if present, will transmit the second and third letters of the localizer ID.

Q. The glide path projection angle for an ILS is usually between _____ and _____°.
A. 2 - 3.

Q. The glide path may not be suitable for navigation below the published Decision Height, therefore any reference to the glide path below the Decision Height must be supplemented by what?
A. Visual contact with the runway surface or approach lights.

Q. How is the localizer signal (frequency) identified on the approach chart?
A. It will always be preceded by the letter "I".

Q. What frequency range is reserved for localizer frequencies?
A. 108.1 - 111.95 MHz (odd-tenths only).

Q. What are the reliable reception limits for a localizer?
A. - 10 NM arc: 35° either side of centerline.
- 10-18 NM arc: 10° either side of centerline at 1000-4500′ agl.

Q. Is it possible for an aircraft to receive glide slope indications while on a Back Course approach?
A. Yes, but the signal is not reliable and should be ignored.

Q. Name seven types of nonprecision approaches.
A. (1) VOR - an approach using a VOR
(2) NDB - an approach using an NDB as the transmitter and an ADF as the receiver.

(3) LOC - an approach using a localizer (ILS minus glide slope information)
(4) LOC BC - an approach using the back course of a localizer
(5) ASR - airport surveillance radar (a ground controlled approach)
(6) SDF - simplified directional facility
(7) LDA - localizer-type directional aid

Q. Both SDF and LDA approaches are similar to the standard localizer approach. In what ways do they differ from the localizer approach?
A. An LDA course is not aligned with the runway; an SDF may or may not be aligned with the runway. Both courses are wider than the standard localizer course, resulting in less precision. The LDA course width is between 3° and 6° and thus provides a more precise approach course than a similar SDF installation which may have a course width of 6° or 12°. Circling minimums normally apply to LDA approaches but may have published straight in minimums if the angle between the runway centerline and the LDA course does not exceed 30°.

NOTE: LDA frequencies are preceded by the letter "I" just like localizer frequencies.

Q. Given the figure below, answer the following questions:

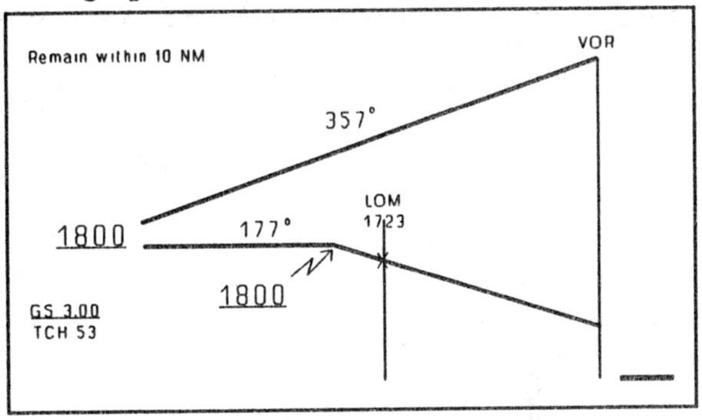

(1) What does the line under 1800 signify?
(2) If you saw an altitude with a line **over** it what would it mean? A line **under** and **over** it? An altitude with no lines above or below it?
(3) Would you expect the profile below to be associated with a precision or nonprecision approach? How can you tell?
(4) What does the "lightning bolt" next to 1800 mean?
(5) The "X" in this figure under the Locator Outer Marker (LOM) represents a maltese symbol. What significance does this symbol have?
(6) What does "GS 3.00°" mean and how can this information be useful to you?
(7) What does "TCH 53" mean?
(8) What is the purpose of the number "1723" under the LOM?
(9) The vertical line passing through both the VOR and the LOM is solid. Many profile views will have vertical dashed lines. What do these mean?
(10) Is this a **Procedure** or **Non-procedure** approach?
A. (1) This represents the **minimum** altitude authorized for that portion of the approach. In this case, it represents the **minimum** altitude that must be maintained for the course reversal procedure (**within 10 NM**) and until intercepting the glide slope on the final approach segment.
(2) An altitude with a line over top of it indicates a **maximum** altitude. An altitude with a line under and over it indicates a **mandatory** altitude (you should not go above or below this altitude). No lines indicates a **recommended** altitude.
(3) This profile is associated with a precision approach (ILS), but has nonprecision contingencies (LOC). There are several ways you can tell. One is

the presence of the "lightning bolt" and another is the presence of glideslope angle information ("GS 3.00°"). Also, the number "1723" under the LOM would not be presented on a strictly nonprecision approach.
(4) The "lightning bolt" represents the minimum altitude for glide slope interception and "points" to the approximate location where the glide slope will be intercepted at that altitude. In the case of a nonprecision approach (ie, LOC), the altitude represents the minimum altitude until crossing the outer marker (LOM).
(5) The maltese symbol represents the Final Approach Fix (FAF) for a nonprecision approach (ie, LOC). In this case, if you were executing the localizer approach you would be required to start your clock here in order to determine the MAP. In the case of a precision approach (ILS), the FAF is the point where the glide slope is intercepted when at the "lightning bolt" altitude. This may be at or before the maltese symbol. In this case it is before.
(6) This indicates that the glide slope has a descent angle of 3°. Using the RATE OF CLIMB/DESCENT chart inside the back cover of any approach volume you can see that a glideslope of 3° and a ground speed of 90 knots equates to a rate of descent of 450 fpm. Once the glide slope is intercepted, a descent of 450 fpm (as indicated on the VSI) should keep you on the glide path (assuming that a ground speed of 90 knots is maintained).
(7) For practical purposes, the TCH represents the height above the runway threshold that an aircraft's receiving antenna should be (it does not represent the height of the glide slope above the runway threshold).
(8) This number represents the altitude of the glide slope over the LOM. After intercepting the glide slope at the "lightning bolt" altitude you should begin a descent to remain on the glide slope. When your altimeter shows this altitude (1723), you will be directly over the LOM. In this example you are flying inbound on the final approach at 1800' msl. You intercept the glideslope at some point short of the LOM and begin a descent. At 1723' msl you will be directly over the LOM. NOTE: As mentioned earlier, it is a good idea to start your clock here. If you happen to lose the glide slope you can revert to a localizer approach and you will have an accurate time for determining the MAP.
(9) A solid line is used to represent facilities such as NDB, VOR, VORTAC, TACAN, WPT, etc. A dashed line is used to indicate a fix or intersection.
(10) Procedure. A "non-procedure" approach would be one where the missed approach point is located at or beyond the end of the approach runway. The vast majority of approaches are procedure approaches. A procedure approach with a missed approach point directly over the airfield is often referred to as a "terminal" approach.

Q. Question "(6)" above refers to a glide slope angle of 3.00°. The answer suggests the use of the RATE OF CLIMB/DESCENT chart inside the back cover of approach volumes. What other method can be used to quickly determine the rate of descent necessary to remain on this glide slope?
A. Using the E6-B, consider that the desired ground speed and the rate of descent in fpm are read on the outer scale. If you plan to execute the approach at 90 kts, place the black arrow or the "hour index" on the inner scale under the number "90" on the outer scale. Read around the inner scale to the published glide slope angle (where 3.00° = 30). Above "30" on the inner scale is the

number "45" on the outer scale. This represents a 450 fpm rate of descent to stay on the glide path.

Q. Sometimes the glide path will have a large **"V"** on it. What does this mean?
A. This represents the Visual Descent Point (VDP).

Q. Given the figure below, answer the following questions:

CATEGORY	A	B	C	D
S-ILS 18	329 - 1/2 200 (200-1/2)			
S-LOC 18	500 - 1/2 371 (400-1/2)			500 - 3/4 371 (400-3/4)
CIRCLING	600 - 1 469 (500-1)	600 - 1 1/2 469 (500-1 1/2)		700 - 2 569 (600-2)

When tower is closed use Hartford altimeter setting and increase all MDAs and DHs 140 feet

▼ T

▲ A NA

(1) Enroute to this airfield you tune up the ATIS frequency and hear the following information: "Somewhere airfield, information Juliet, measured 400 overcast, ¼-mile visibility, winds 270 at 10 gusting to 15. ILS runway 18 in use, all aircraft landing and departing runway 28, altimeter setting 29.94, advise on initial contact you have information Juliet." You are told to expect vectors for the ILS RWY 18. Can you legally shoot this approach? If so, what will be your Decision Height? Would you expect to encounter VMC conditions at this altitude?
(2) If you were shooting the straight-in ILS RWY 18 approach, what would be your Decision Height? How far above the airport is this?
(3) What kind of weather would you need to plan a flight to this airfield?
(4) What kind of weather would you need to plan a flight to this airfield and **not** need an alternate (control zone is in effect, radar is not required for the approach, and all navaids are monitored). Can I use this airfield as an alternate?
(5) If you lose the glide slope can the approach be continued? What would be your Decision Height? What happens if you lose the localizer signal?
(6) What weather would be required to file for the ILS RWY 18 circle to land RWY 28 at 100 kias?
(7) You are planning the ILS RWY 18 approach to this airfield and notice that there is a black star next to the control tower frequency in the Plan View. After checking the IFR Supplement you realize that the tower will be closed at your expected time of arrival. How does this affect your Decision Height? Does this change the landing weather minimums?
A. (1) You **can** legally shoot this approach since AR 95-1 states, "An approach may be initiated, regardless of ceiling and visibility." However, since all aircraft are landing and departing runway 28 you would have to use circling minimums instead of the straight-in minimums. Here you would not be authorized to descend below 600'msl (MDA) unless you had the landing area or runway lights in sight. Furthermore, you can see that the height of the MDA above the airport is 469'agl. If the ceiling is at 400'agl as reported, chances are you would not encounter VMC conditions. NOTE: This situation is not likely to occur if IMC conditions exist anywhere near the MDA. The purpose of this question is to test your understanding of circling minimums. Practice approaches in VMC conditions are more likely to result in the use of such minimums. Remember that a straight-in ILS has a DH, whereas a circle-to-land ILS has an MDA.

(2) 329′ msl; circling minimums will include a Height Above Airport (HAA) figure, all others will have a Height Above Touchdown Zone (HAT) figure. In this case, the Decision Height will be 200′ above the Touchdown Zone Elevation (TDZE). Since elevations around the airport may vary this may not accurately reflect the actual height above the airport.
(3) Weather must be forecast to be equal to or greater than 200 - ¼ at ETA through one hour after ETA.
(4) Worst weather forecast for this field is equal to or greater than 600 - 1¼ at ETA through one hour after ETA. The "A-NA" (alternate not authorized) in the NOTES section indicates that I may not use this airfield as an alternate unless a descent from minimum enroute altitude for IFR operation can be made in VFR conditions.
(5) Yes, assuming you started your time at the FAF (maltese). Since a localizer approach is considered a nonprecision approach you would not have a Decision Height (DH) but a Minimum Descent Altitude (MDA). The MDA would be 500′ msl and the MAP could be determined using the FAF to MAP Table in the Airport Data Section of the chart (lower right corner). If the localizer signal is lost then the approach must be discontinued.
(6) AR 95-1 states, "Aviators flying helicopters may reduce destination and alternate Category **A** minimums by 50 percent, but not less than 1/4 mile or metric equivalent." Shooting the approach at 100 knots indicated airspeed places you in Category B minimums. In this case, since you cannot reduce Category **B** minimums you would need at least a 600′ ceiling and 1-½ miles visibility.
(7) The NOTES section indicates that when the tower is closed you are to use Hartford altimeter settings and increase all MDAs and DHs by 140 feet. Thus, your new DH will be 329+140 or 469′ msl. The weather planning minimums must also be recomputed whenever the MDA or DH is raised due to the use of a remote altimeter setting. In some cases the new weather minimums are shown in the minimums box; otherwise, they must be computed manually. Use the following formula:

  MDA/DH with local altimeter
+ Increase when using remote altimeter
= New MDA/DH
- Airport elevation
= New weather planning minima (rounded up)

The airport elevation will be shown in the airport data section or it may be computed by subtracting the Height Above Airport (HAA) from the circling MDA. In this case, the elevation is 600-469 or 131′agl. Plug in the known variables:

DH 329
+ 140
= 469
- 131
= 338 (rounded up = 400 - 1)

400 - 1 is the new weather planning minima which, of course, may be reduced to 400 - ½.

Q. What is an "HAL" and when is it used?
A. Height Above Landing area; used in COPTER minima only.

Q. What is an "HAS" and when is it used?
A. Height Above Surface; this is the height of the MDA above the highest terrain/surface within a 5200' radius of the MAP in Point in Space procedures (helicopter only procedures).

Q. Are circling minimums ever provided for COPTER ONLY procedures?
A. No.

Q. Ceilings are always reported in feet ________ (msl/agl) and prevailing visibilities are reported in _________ __________ (statute miles/nautical miles).
A. AGL; statute miles (and fractions thereof).

Q. What does RVR stand for, how is it expressed and how is it measured?
A. Runway Visual Range; expressed in 100s of feet; represents distance measured down a runway using a transmissiometer.

Q. How is an airport beacon represented on an airport diagram?
A. By a black star with a white circle in the middle.

Q. How is an airport's control tower represented on an airport diagram?
A. By a small black square "■".

NOTE: Black squares larger than the control tower represent buildings in the vicinity of the airport.

Q. When an airport control tower and the beacon are co-located, how is this represented?
A. The beacon symbol is used with the letters "TWR" next to it.

Q. What is a "GCA"?
A. Ground Controlled Approach. This is a radar approach.

Q. What are the two types of ground controlled approaches?
A. ASR and PAR.

Q. You would expect lower minimums for which of the above approaches?
A. PAR, this is a precision approach.

Q. How do you determine the availability of a radar approach at an airfield?
A. If a radar approach is available you will see the letters "ASR", "PAR", or "ASR/PAR" under the communications data in the Plan View section of the approach chart.

Q. What do "PAR" and "ASR" stand for?
A. Precision Approach Radar; Airport Surveillance Radar.

Q. What information is provided to a GCA controller for an ASR approach? A PAR approach?
A. Range (distance) and azimuth (direction); range, azimuth, and altitude (glide

slope).

Q. Who would a pilot call to request a radar approach?
A. Approach control.

Q. Name some methods that can be used by an ASR controller to identify aircraft on his/her scope.
A. - Transponder
- Heading change of 30° or more
- position over a known fix or navaid

Q. While being vectored to final on a radar approach, you should initiate lost communications procedures if you do not receive a transmission from the controller for a period of ________.
A. 1 minute.

Q. While on a radar **final** approach you should initiate lost communications procedures if no transmissions are received for a period of ________ for ASR or ________ for PAR approaches.
A. 15 seconds; 5 seconds.

Q. What are the lost communications procedures for radar approaches?
A. - Attempt contact on a secondary frequency, the previously assigned frequency, the tower frequency or guard.
- If unable to reestablish communications and unable to maintain VMC, proceed with a published instrument approach procedure or previously coordinated instructions. Change transponder to appropriate codes (7600).
- Maintain the last assigned altitude or the minimum safe/sector altitude (emergency safe altitude if more than 25 NM from the facility), whichever is higher, until established on a segment of the published approach.

Q. How do you know what the missed approach procedures are for a radar approach?
A. If the procedure is not published the final controller will give you this information.

Q. ASR scopes can show the controller the azimuth and range of the aircraft within ______° of the airport. How does this compare to azimuth and range information for PAR scopes?
A. 360°; PAR scopes have a 20° "sweep" and show information that is 10° right or left of runway centerline (out to a range of 10 NM).

Q. While an aircraft is on PAR final, the controller should give a "distance from touchdown" advisory to the pilot at least how often?
A. At least once each nautical mile.

Q. Is it the pilot's or the controller's responsibility to initiate missed approach procedures if VMC conditions are not reached at the DH/MDA?
A. The pilot's.

Q. Where would you look to determine radar approach minimums?
A. Radar minima can often be found in the LANDING MINIMA section of an airfield that has such capabilities. More detailed information can be found in the RADAR INSTRUMENT APPROACH MINIMUMS section of the volume just after the non-standard departure tabulation.

Q. Given the information listed below (a portion of a RADAR INSTRUMENT APPROACH MINIMUMS table), answer the following questions:
(1) What is the DH for a PAR RWY 6 approach (CAT A)?
(2) What is the MDA for an ASR RWY 36 approach?
(3) What is the MDA for an ASR RWY 6, circle-to-land RWY 18 approach?
(4) What are the weather minima for an ASR RWY 36, circle-to-land RWY 24?

	RWY	CAT	DH/MDA-VIS	CEIL-VIS
PAR	6	AB	498/24	(200-½)
		CD	498/40	(200-3/4)
	36	ABCD	496-3/4	(200-3/4)
ASR	6	ABC	680/40	(400-3/4)
		D	680/50	(400-1)
	36	ABC	680-1	(400-1)
		D	680-1¼	(400-1¼)
CIR	6-24	ABC	740-1	(500-1)
		D	760-1	(500-1)
	18-36	AB	720-1	(500-3/4)
		CD	740-1½	(500-1)

A.
(1) 498′
(2) 680′
(3) 740′. In the case of a ground controlled approach, circling MDA and weather minima to be used are those for the runway to which the final approach is flown - not the landing runway. In this example, even though you are circling to land to RWY 18, the MDA to be used must be the RWY 6 circling MDA.
(4) 500-3/4

Q. If RVR minimums for take-off or landing are published in an instrument approach procedure, but RVR is inoperative and cannot be reported for the runway at that time, then it will be necessary to convert to ground visibility. Where would you look to make this conversion? An RVR of 50 would be equivalent to how many miles ground visibility?
A. A METER CONVERSION CHART is provided in the legend information at the front of each volume; 1 mile.

Q. What is a "lead radial" and how is it represented on the approach chart?
A. A lead radial, when published, represents the point at which the turn inbound on the localizer course begins. It is represented by the letters "LR" (ie, LR-140).

Q. May an Army aviator use dual VOR minimums (if available) and the associated weather planning minimums published on approach charts?

A. Yes, using a single VOR and off-tuning from the approach aid to identify an approach fix is authorized for Army aircraft.

Q. What does the acronym "UNMTO" represent?
A. Unmonitored.

Q. A VFR sectional chart must be used for planning what portion of an IFR flight?
A. Direct routes of flight (used to determine minimum IFR altitude (MIA)).

Q. What is the procedure for determining the appropriate IFR altitude for direct legs of flight (off airways or feeder routes)?
A. As mentioned in the above question, grab a VFR sectional covering the direct leg concerned. Draw a course line. Determine the highest obstacle within 4 NM either side of your course line for the entire leg. Add 1,000′ (non-mountainous areas) or 2,000′ (mountainous areas) and round up as necessary to arrive at an appropriate altitude using the semi-circular rule.

For example, let's say that the highest obstacle, as determined above, were 1,700′ msl. In a non-mountainous area you would add 1,000′. This gives you 2,700′. If your direction of flight is toward the west then by the semi-circular rule you would need to file for a minimum altitude of 4,000′ msl for the entire leg.

Q. AR 95-1 mentions that any airfield may be used as an alternate if you are able to descend from the enroute minimum altitude for IFR operations, as well as make the approach and landing, in VFR conditions. How do you determine this?
A. If the final leg of the approach will be on an VICTOR AIRWAY, use the MEA (or MOCA if there is one). Subtract from this the field elevation of the destination airport to give you an altitude "above ground level" or AGL. Add 500′ to this number (to meet cloud clearance requirements). If the predominant weather at the destination airfield at ETA through one hour after ETA is at or above this altitude then the necessary criteria have been met.

If the final leg of the approach will be a DIRECT ROUTE, the minimum altitude for IFR operations is determined as explained in the previous question. From this value subtract the destination field elevation and add 500′ as above.

NOTE 1: A published FEEDER ROUTE may provide an altitude that is lower than an MEA or MOCA along the enroute victor airway. There is some debate, however, as to whether or not a feeder route altitude may be used since it is "technically" a portion of the <u>initial approach</u> segment and not the <u>enroute</u> segment.
NOTE 2: The whole purpose of these calculations is to answer the following question: "What is the lowest the clouds can be at my destination that will allow me to fly at my minimum IFR altitude and still maintain cloud clearance requirements (VFR conditions)?"

Basic Cloud Clearance for Class E airspace means that I must be able to remain 500′ or more below any clouds in the vicinity. Since cloud levels are reported in feet AGL instead of MSL a conversion becomes necessary.

Remember that "cloud clearance" means just that - clearance from all clouds. Don't use a reported ceiling for calculations, use the lowest reported layer of clouds.

GENERAL AND INSTRUMENT FLIGHT RULES

In Flight School the block of instruction on "General and Instrument Flight Rules for Rotary Wing Aircraft" deals primarily with AR 95-1, "Flight Regulations". For this reason it is presented first in this chapter. Following the section on AR 95-1 are sections dealing with Publication Cycles, GP, AP, FIH, and Enroute IFR Supplement and is presented for those aviators wishing a more in-depth study or review.

AR 95-1

Q. AR 95-1 establishes responsibilities, procedures, and rules for aircrew training, standardization, and the operation of Army aircraft. What is the only time that you are authorized to deviate from this regulation?
A. During emergencies.

Q. Individuals who deviate from the provisions of AR 95-1, Federal Aviation Administration (FAA) or host country regulations must report details of the incident direct to their unit commanders within _______ hours after it occurs.
A. 24

Q. Violations of Federal Aviation Regulations (FAR), International Civil Aviation Organization (ICAO), host country, and military aviation regulations will be treated as outlined in what other regulation?
A. AR 95-3.

Q. What may be the result of violations reported IAW AR 95-3?
A. May result in administrative, nonjudicial or judicial actions IAW AR 15-6 and AR 600-105.

Q. An aviator is prohibited from performing aircrew duties if attending a nonflying course of instruction having a duration of more than _____ days.
A. 90

Q. In addition to aviators that fall into the above category, name two other categories of aviators that are prohibited from performing aircrew duty.
A. - Commissioned officers in non-operational aviation positions.
- Those aviators disqualified, temporarily suspended, or whose aviation service is administratively terminated IAW AR 600-105.

Q. What publications and forms are required to be in the aircraft prior to flying?
A. The exact publications and forms required prior to flight will vary slightly from station to station. However, the following should always be onboard the aircraft:

- Operator's manual (-10) and -10 checklist
- DD Form 1896 (Jet Fuel Identaplate)
- All publications necessary for flight (approach plates, etc)
- Aircraft logbook
    - DA Form 2408-31, Aircraft Identification Card
    - DA Form 2408

- DA Form 2408-12, -13, -14
- HIT check log
- PMD/PMS inspection checklist
- DD Forms 365-4 (or computer generated equivalent) in sufficient quantities to complete mission
- Any applicable Airworthiness Releases

Q. According to AR 95-1, when are checklists required to be used?
A. From before starting engines through shutdown, on maintenance operational checks (MOCs), maintenance test flights (MTFs), and daily inspections.

NOTE: While airborne, when time may not permit use of the checklist or when its use would cause a hazard to safety, required checks may be from memory.

Q. What stipulation is made concerning the use of publications and checklists on aircraft?
A. Only DA approved operator's manuals and checklists will be used.

Q. When logging aircraft or simulator flight time an entry must be made on DA Form 2408-12 (Army Aviator's Flight Record) for each crewmember indicating the duties performed, mission, and flight condition. Identify the following duty symbols: PC, PI, CP, IP, IE, SP, CE, MO, OR, MP, UT.
a. PC - Pilot in Command
PI - Pilot
CP - Copilot
IP - Instructor Pilot
IE - Instrument Examiner
SP - Standardization Instructor Pilot
CE - Crew Chief or aircraft mechanic assigned to crew chief position
MO - Flight surgeon or other medical personnel
OR - Aircraft maintenance personnel, technical observer, gunner, or other duties requiring flight.
MP - Maintenance test Pilot
UT - Unit Trainer

Q. When may an aviator log time using the "CP" symbol?
A. If either of the following is true:
- If at the copilot station (access to controls) but not qualified or current in the aircraft being flown, or
- If performing copilot duties at other than the copilot station and undergoing training or evaluations conducted by an IP, SP, IE, UT, or ME.

Q. What are some examples of copilot duties when at other than the copilot station?
A. This can include practically any of the normal copilot duties not involving the actual flying of the aircraft (ie, navigation, communication, fuel checks, determine proper procedure for entering holding, etc).

Q. Can an instructor pilot log flight time as a "PC"?

A. Not if performing assigned duties as an IP (which they normally do). The specific symbol will be used to indicate the duty being performed by the pilot in command (ie, IP, IE, UT, SP, MP, etc).

Q. When the UT symbol is used, what must be completed for the aviator undergoing training?
A. A gradeslip.

NOTE: Only one aviator occupying a flight crew station may use any one symbol for any time period.

Q. Identify the following mission symbols: S, T, F, C, D.
A. S - Service missions (other than A, C, F, T, X or D).
T - Training flight for individual qualification, refresher, mission, or continuation.
F - Maintenance test flight
C - Combat mission directly against the enemy within a designated combat zone.
D - Imminent danger (applies when Imminent Danger Special Pay is authorized).

Q. Identify the following flight condition symbols: D, N, NG, W, H, TR.
A. D - Day
N - Night
NG - Night Goggles (night vision goggles used during night flight).
W - Weather
H - Hood
TR - Terrain

NOTE: DA Form 2408 specifies that "NG" will be logged when using AN/PVS-5 while "NV" will be logged when using AN/AVS-6 (ANVIS). Additionally, "NS" will be logged when using Night Vision Systems (ie, AH-64).

Q. When can an aviator officially log "night" flight time (N/NG/NV/NS)?
A. Between the hours of official sunset and official sunrise.

Q. When logging flight time as either "H" or "W", the aircraft attitude must be controlled using aircraft instruments. What is the difference between them?
A. When logging flight time as "W", you must be in actual weather conditions that do not permit visual contact with the horizon or earth's surface; when logging flight time as "H", the vision of the person flying the aircraft is artificially limited from viewing the horizon or earth's surface.

NOTE: The flight condition symbol "W" is also used to log instrument time flown in a flight simulator.

Q. For the purpose of logging flight time, when does flying time begin and end?
A. Flying time starts when an airplane begins to move forward on the takeoff roll or when a helicopter lifts off the ground. Flying time ends when the aircraft has landed and the engines are stopped or the flying crew changes.

Q. When assigned to a new unit, all crewmembers must submit their individual flight records to the unit within ______ calendar days after reporting for duty.
A. 14

Q. Give some examples of flights that are restricted to minimum essential personnel.
A. - Maintenance test flights
- Aerial demonstrations
- Flights involving emergency procedures training, NVD training, and refresher training per the ATM.
- Hazardous flight training
- Aircraft acceptance flights

Q. Smoking is prohibited within _______ feet of an aircraft on the ground.
A. 50

NOTE: Smoking is prohibited in aircraft (on ground or in flight).

Q. Procedures for packaging, handling, and air transportation of dangerous materials are prescribed in what regulations?
A. AR 95-27 and AR 55-203.

Q. What form is used to notify commanders and safety councils of anything affecting the safety of Army aircraft or related personnel and equipment?
A. DA Form 2696-R (Occupational Hazard Report - OHR).

Q. In unpressurized aircraft, when is oxygen required to be used?
A. (1) Crew - flights above 10,000' for more than 1 hour;
- flights above 12,000' for more than 30 minutes;
(2) Crew and passengers - above 14,000' for any period of time.

NOTE: For flights above 18,000' oxygen prebreathing will be accomplished by aircrewmembers.

Q. AR 95-1 lists six items of approved clothing and equipment that must be worn by all crewmembers while performing crew duties. What are they?
A. (1) Leather boots
(2) Flight helmet
(3) Flight suit
(4) Flight gloves
(5) Cotton, wool, or nomex underwear
(6) Identification tags

Q. A pilot in command must ensure that there are installed seats and seat belts for every passenger on the aircraft and that all passengers can operate the seat belts and, if installed, shoulder harnesses. When are passengers <u>required</u> to wear seat belts and, if installed, shoulder harnesses?
A. During takeoffs, landings, and in the event of turbulence.

Q. When are the pilot and copilot required to wear properly adjusted seat belt and shoulder harness?
A. When at the controls.

Q. MACOM commanders and general officers are authorized to waive the requirements for seats/seatbelts in aircraft under their commands during what activities?
A. - Parachute drops
- Rappelling
- Life or death missions
- Water insertions and extractions

Q. Can commanders authorize nonaviator personnel to start, operate and stop aircraft auxiliary power units (APUs)?
A. Yes, if they are trained on all functions he/she is authorized to perform and have written authorization from the commander.

Q. Army aircrew training must be conducted IAW what publications?
A. The Aircrew Training Program (ATP) and the Aircrew Training Manual (ATM).

Q. If _______ days have elapsed since the last flight as pilot or pilot in command in the aircraft mission, type, design, and series to be flown, the aviator will be administered a Proficiency Flight Evaluation (PFE) per the ATM.
A. 60

Q. What symbols identify the "mission, type, design, and series" of the Huey helicopter?
A. UH-1H;
Mission = U = Utility
Type = H = Helicopter
Design = 1 = One
Series = H = Hotel

NOTE: Currency in any one aircraft will satisfy the requirement for all aircraft <u>in the series</u> (ie, UH1, B, H, V, EH-1H, EH-1X).

Q. DA From 5484-R (Aircrew Mission Briefing), also known as a "mission brief sheet", must be retained in unit files for how long?
A. 30 days

Q. What is required to be designated by the unit commander for flights consisting of two or more aircraft operating as a single flight?
A. An Air Mission Commander (AMC).

Q. In most cases, any aviator can occupy a copilot station and perform designated copilot duties in any aircraft (see NOTE below). One exception is for flight into forecast instrument meteorological conditions (IMC). Explain this exception.
A. Two aviators current in the aircraft <u>category</u> being flown are required for flights into forecast IMC.

NOTE: This question pertains to aircraft having the requirement for only one pilot (for instance, the UH-1). In such cases, the other station is the copilot station.

Q. What is meant by "category" in the above question?
A. Category means aircraft designated as either airplane or helicopter.

Q. How many hours of pilot time in an aircraft category are desirable for an individual to become an instructor pilot?
A. 500

Q. Can an aviator become an IP without going to the IP course?
A. Yes. In rare cases a commander may coordinate with DES, USAAVNC and submit a request through HQDA for an IP equivalency evaluation to be given to an aviator.

Q. What are considered to be the primary references governing the operation of a specific aircraft?
A. The aircraft operator's manual (commonly referred to as the -10) and checklists.

Q. When differences exist between AR 95-1 and other publications, which has precedence?
A. AR 95-1

Q. What action is required to be taken for aviators involved in a Class A or B mishap? A Class C mishap?
A. Aviators with access to flight controls and performing crew duties involved in a Class A or B mishap <u>will</u> be suspended from flight duties until successful completion of a flight evaluation. Aviators with access to flight controls and performing crew duties involved in a Class C mishap <u>may</u> be suspended from flight duties and required to successfully complete a flight evaluation at the discretion of the commander.

Q. What does AR 95-1 mention as required for preflight planning?
A. An evaluation of aircraft performance, approach and departure data, notices to airmen (NOTAM), and the Army Aviation Flight Information Bulletin.

Q. What fuel reserve requirements should be planned for VFR/IFR flights?
A. At takeoff, aircraft must have enough fuel to reach the destination and alternate airport (if required) and have a planned fuel reserve of 20 minutes (at cruise) for VFR flights and 30 minutes (at cruise) for IFR flights.

Q. What restrictions are stated about flight into icing conditions?
A. - Aircraft <u>will not</u> be flown into <u>known or forecast severe</u> icing conditions.
- If a flight is to be made into <u>known or forecast moderate</u> icing conditions, the aircraft must be equipped with adequate deicing or anti-icing equipment.

Q. What restrictions are stated about flight into turbulence?

A. - Aircraft will not be flown into known or forecast extreme turbulence or into known severe turbulence.

- Aircraft will not be flown into forecast severe turbulence unless MACOM commanders have established clearance procedures and other restrictions, as specified in AR 95-1, are met.

Q. What restrictions are stated about flight into thunderstorms?
A. Intentional flight into thunderstorms is prohibited.

Q. What are the general weather requirements for planning a VFR flight?
A. Destination weather must be forecast to be equal to or greater than VFR minimums at ETA through 1 hour after ETA. When there are intermittent weather conditions, predominant weather will apply.

Q. Can you file VFR to a Class B, C, or D airfield if the destination conditions are IFR (weather less than 1000 - 3)?
A. If the known conditions are at or above special VFR (SVFR) minimums for that airfield at ETA through 1 hour after ETA, then aviators may file flights to such destinations (this also includes surface-based Class E airspace).

Keep in mind that a special VFR clearance must be obtained from the controlling agency (approach, tower, etc) and an alternate airfield should be planned in the event that a SVFR clearance cannot be obtained.

NOTE: Outside of controlled airspace, forecast enroute weather must permit flight with separation from clouds and flight visibility equal to or greater than minimums stated in AR 95-1, table 5-1 (see next two questions).

Q. What are the rotary wing VFR weather minimums for uncontrolled airspace?
A. Day - BCC, ½ mile visibility
Night - BCC, 1 mile visibility

BCC = Basic Cloud Clearance = 500' below
1,000' above
2,000' horizontally from clouds

Q. What is the exception to the above VFR minimums?
A. For helicopters operating at or below 1200' agl, regardless of msl altitude, cloud clearance minimums change to clear of clouds so long as the helicopter is operated at a speed that allows the pilot adequate opportunity to see any air traffic or obstruction in time to avoid a collision (airspeed to "see and avoid"). No change to visibility requirements ==> ½-mile day; 1-mile night.

Q. What are the general weather requirements for planning an IFR flight?
A. Destination weather must be forecast to be equal to or greater than the published weather planning minimum for the approach procedure to be flown at ETA through 1 hour after ETA. When there are intermittent weather conditions, predominant weather will apply.

Q. Aviators flying helicopters may reduce destination and alternate Category A

visibility minimums by ______%, but not less than ______ mile or metric equivalent.
A. 50%; ¼.

Q. Does the above rule apply to approaches labeled "COPTER ONLY"?
A. No, reduction of visibility for approaches labeled "COPTER ONLY" is not authorized.

Q. Can Category II approach procedures be used in destination or alternate weather planning?
A. No.

Q. How long is a weather forecast good for once received?
A. 1 hour and 30 minutes from the time the forecast was received <u>provided the aircraft to which it applies has not departed</u>.

Q. A military weather briefing should be completed on what form?
A. DD Form 175-1 (Flight Weather Briefing), often referred to as simply a "-1".

Q. Who has flight plan approval authority (signs in SIGNATURE OF APPROVAL AUTHORITY on DD Form 175, Military Flight Plan)?
A. The Pilot in Command for aircraft under his/her control when performing missions directed by the commander.

Q. AR 95-1 states that all Army aircraft that are instrumented for IFR flight and are flown by an instrument rated pilot will operate on an IFR flight plan. Some exceptions are listed. What are they?
A. - If flight is primarily for VFR training.
- If time will not permit mission completion under IFR.
- If mission can only be accomplished under VFR.
- If excessive ATC departure, enroute, or terminal area delays are encountered.
- If hazardous weather conditions must be avoided.
- If two instrument rated pilots current in aircraft category being flown are not at the controls (or an IP/IE at one set of controls).

Q. Stopover flight plans are subject to what two provisions?
A. - No change should be made in pilot in command.
- If the original manifest does not list passenger or crew changes at stopover points, changes will be filed with military installation base operations, FAA flight service, or other competent authority.

Q. When departing a nonmilitary airfield what information must be passed to a Flight Service Station as soon as possible?
A. The exact time of departure from the nonmilitary airfield (UTC time).

Q. A crew and passenger manifest is required for all flights. How should this be accomplished for the tactical transport of troops (air assault) or tactical training flights?

A. The passenger manifest should be prepared and retained by the supported unit.

Q. What weather planning questions should be asked when planning an IFR flight?
A. - "What weather do I need to legally file to my destination?"
- "What weather do I need in order to not need an alternate?"
- "Does my alternate selection, if needed, qualify?"

Q. Under what conditions would an alternate airfield be required when filing IFR to a destination?
A. If any of the following are true:
- Radar is required to execute the approach procedure to be flown.
- Navaids for the instrument approach to be used are unmonitored.
- The predominant weather at the destination is forecast to be less than 400-1 above the weather planning minimum required for the approach to be flown at ETA through 1 hour after ETA.

NOTE: The one exception to the above is that, regardless of the status of radar or navaids, an airfield does not require an alternate if descent from enroute minimum altitude for IFR operation, approach, and landing can be made in VFR conditions.

Q. Careful planning should be involved in the selection of an alternate airfield. What questions should be asked when deciding whether or not an airfield is acceptable for use as an alternate?
A. 1. What is the worst weather condition forecast for the airfield at ETA through 1 hour after ETA? **If you can descend from enroute minimum altitude for IFR operation as well as make the approach and landing in VFR conditions then the airfield may be used as an alternate regardless of the answers to the next four questions.** If the worst weather conditions are less than 400-1 above the weather planning minimum required for the approach to be flown then the airfield may not be used as an alternate.

2. Is the approach procedure to be used at the alternate shown as not authorized (NA) in FLIP? If the answer is YES then the airfield may not be used as an alternate unless the conditions in #1 above apply.

3. Is radar required for the approach procedure to be used at the alternate? If the answer is YES then the airfield may not be used as an alternate unless the conditions in #1 above apply.

4. Are the navaids to be used for the approach procedure at the alternate airfield unmonitored? If the answer is YES, that is, if the navaids are listed in FLIP or NOTAMed as "unmto" then the airfield may not be used as an alternate unless the conditions in #1 above apply.

5. Is a control zone in effect at the alternate? If a control zone does not exist or is not in effect at the airfield then it may not be used as an alternate unless the conditions in #1 above apply.

MEMORY AID: For determining if airfield is suitable as an alternate:

W - weather 400-1 above planning minima
A - no "A(NA)"

R - radar not required
N - navaids monitored
C - control zone in effect

Use the same acronym minus the "A" and "C" for determining if an alternate is required.

W - weather less than 400-1 above planning minima
R - radar required for approach to be flown
N - navaids to be used for approach are unmonitored

Q. What does AR 95-1 specify that a pilot must ensure with regards to aircraft weight & balance.
A. The pilot must ensure that the aircraft is within weight and center-of-gravity limitations for the duration of the flight and that sufficiently completed and accurate DD Forms 365-4 are on board the aircraft to verify this information.

Q. How often must a new weight & balance sheet be recomputed for aircraft?
A. Must be recomputed <u>or verified</u> within the previous 90 days (1 year for USAAVNC, per USAAVNC SOP).

Q. In what publication would you look for information regarding aircraft weight and balance?
A. AR 95-3, Chpt 6.

Q. When should the aircraft anti-collision light be on?
A. Whenever the aircraft engines are operating.

Q. What are the exceptions to the above question?
A. - When conducting night vision device operations (in designated "blackout" areas).
- When conditions may cause vertigo (ie, in a cloud bank).
- When there may be other hazards to safety.

Q. When should aircraft position lights be on BRIGHT?
A. Between official sunset and official sunrise.

Q. What are your IFR takeoff minimums?
A. - If you have logged more than 50 hours of actual weather time as a PC, UT, IP, SP, IE, MP, ME, XP, or pilot (see NOTE) then you do not have any takeoff minimums; in other words you could conceivably, and legally, take off in 0/0 conditions.
- If you have not logged at least 50 hours of actual weather time as per above then your takeoff minimums are 100-¼ (100′ ceiling and ¼ mile visibility, 1200′ RVR or metric equivalent).

NOTE: The term "pilot" above is taken to mean that weather time logged as "Pilot" in Part II of DA Form 759. This portion of DA Form 759 is a historical record of hours flown prior to the initiation of the form.

The reason for this section is that prior to this form, "pilot-in-command" time was logged as "pilot" and all other time was logged as "copilot." Even though "PI" time is, in effect, "pilot" time, weather time logged as "PI" does not count toward the 50-hour requirement mentioned above.

Q. Can flight time logged as "W" in a simulator count towards the 50 hours of weather time (see above question)?
A. No.

Q. You have 20 hours of actual weather time and you are with an IE that has logged over 100 hours of actual weather time. What are your IFR takeoff minimums?
A. This depends on who is at the controls during takeoff. The IE would not have any minimums but you would still have minimums of 100-¼.

Q. What rule applies to the use of Runway Visual Range (RVR) for determining if IFR takeoff minimums have been met?
A. The RVR must be reported for the actual runway used for departure.

NOTE: Be suspicious of a takeoff clearance that specifies a helipad when visibility conditions are only reported using an RVR.

Q. A negative symbology "T" enclosed in an inverted triangle and printed on instrument approach plates indicates that nonstandard IFR takeoff minimums and/or departure procedures apply. All aviators must comply with these when published. When may a pilot in command deviate from these?
A. If approval is received from the agency controlling the departure.

Q. When are SVFR flights within and departures from control zones authorized?
A. Only after an ATC clearance has been obtained.

NOTE: ATC cannot "recommend" a SVFR clearance to enter or depart a control zone. This must be specifically requested by the pilot.

Q. What is the limitation for flying "VFR ON TOP"?
A. 30 minutes unless IMC equipped and IFR requirements can be met for the remainder of the flight.

Q. In addition to continuously monitoring appropriate frequencies while on a VFR flight, how often must position reports be made?
A. Once each hour or as otherwise required.

Q. What weather information is required to be reported by radio?
A. Any unforecast weather and any forecast hazardous weather.

Q. AR 95-2 sets Army policy and procedures for handling airspace matters. In a combat zone however, airspace use, control, and management will be conducted as per what FM?
A. FM 100-103.

Q. What stipulation is made concerning the overflight of National Security Areas?
A. Aviators will avoid overflight of National Security Areas below 2,000' agl (exceptions per DOD FLIP).

Q. When can an aviator request ATC clearance to hold?
A. At any time if fuel and alternate requirements can be met.

Q. An aircraft flying on an IFR flight plan in VMC conditions may be requested by ATC to complete a "visual approach". Is the acceptance of such an approach mandatory?
A. No. Keep in mind, however, that you must have an ATC clearance to conduct a visual approach.

Q. When published landing visibility minimums require conversion between RVR and miles or metric equivalent, the conversion table in DOD FLIP (legend information of approach plates) will be used. Which measurement is the controlling visibility factor when published and reported for a runway?
A. RVR

Q. If published and reported for a runway, can RVR be used with a circling approach?
A. No.

Q. May an Army aviator use dual VOR minimums (if available) and the associated weather planning minimums published on approach charts?
A. Yes, using a single VOR and off-tuning from the approach aid to identify an approach fix is authorized for Army aircraft.

Q. What are the minimum weather requirements necessary to <u>initiate</u> an approach?
A. None. An approach may be initiated, regardless of ceiling and visibility.

Q. Some approach charts have the letters "CAT II" in bold letters at the top and bottom. Are you authorized to make such an approach?
A. Only if shooting the approach in VMC conditions. Category II ILS approaches in IMC are authorized only when all provisions of table 5-3, AR 95-1 are met (very few if any Army helicopters meet these requirements).

Q. Unless otherwise specified in DOD FLIPs, airplanes will fly traffic patterns at _______' agl while helicopters will fly patterns at _______' agl.
A. 1500'; 700'

NOTE: At civilian airfields helicopters will avoid the flow of airplane traffic (unless otherwise instructed by ATC).

Q. An aircraft will not be flown below the published MDA or an approach continued below the DH unless what two conditions exist?
A. - The approach threshold of the runway, or the approach lights or other markings, identifiable with the approach end of the runway or landing area are clearly visible to the pilot.

- The aircraft is in a position from which a safe approach to the runway or landing area can be made.

Q. Explain the semi-circular rule.
A. IFR: for IFR flights in controlled airspace, use an odd thousand altitude if your heading is between 0° and 179° (magnetic); use an even thousand altitude if your heading is between 180° and 359° (magnetic).

VFR: for VFR flights above 3,000' agl, use an odd thousand altitude plus 500' if your heading is between 0° and 179° (magnetic); use an even thousand altitude plus 500' if your heading is between 180° and 359° (magnetic).

MEMORY AID: NEODD - north thru east = odd thousand (plus 500' for VFR)
SWEVEN - south thru west = even thousand (plus 500' for VFR)

Q. Under IFR conditions, no person may operate an aircraft below an altitude of 1000' above the highest obstacle within a horizontal distance of ______ from the course to be flown.
A. 4 NM (this is often referred to as the Minimum IFR Altitude or MIA).

Q. When may an aircraft operate below a published MEA?
A. If a MOCA is published along with an MEA on the route segment being flown, you may operate below the MEA, but not below the MOCA, when within 22 NM of the NAVAID.

The following seven questions are from AR 95-3, Chpt 6, "Weight and Balance."

Q. For weight and balance purposes all Army aircraft are divided into one of two classifications - Class 1 or Class 2. What is the difference between these? What classification does the UH-1 fit into? Where can this information be obtained?
A. Class 1 - those aircraft whose weight or center-of-gravity limits can sometimes be exceeded by loading arrangements normally used in tactical operations. Therefore, limited loading control is needed.

Class 2 - those aircraft whose weight or center-of-gravity limits can readily be exceeded by loading arrangements normally used in tactical operations or those aircraft designed primarily for transporting troops and other passengers. Therefore, a high degree of loading control is needed.

The UH-1 is a Class 2 aircraft.

The weight and balance classification can normally be found in the aircraft operator's manual (-10); however, if the classification is not stated then it should be considered as Class 2.

Q. Sufficient completed DD Forms 365-4 (or computer generated equivalent) must be on board the aircraft to enable the pilot to determine proper aircraft loading for any normal anticipated unit mission and to verify that the weight and center-of-gravity will remain within allowable limits for the duration of a flight. Small changes in the basic weight and moment due to the removal or installation of aircraft equipment or other actions are normally not a cause for concern so long as they are annotated on DD Form 365-3. When should the DD Form 365-4 be

considered invalid with regards to such basic weight and moment changes?
A. When the difference between the entries for basic weight change by +/- 3/10 (three-tenths) of 1%; or when the difference between the entries for center-of-gravity at basic weight change by 0.3 inch.

NOTE: Temporary additions, removals, or relocations of aircraft equipment that are considered part of the basic weight may be entered on DA Form 2408-13 (Aircraft Inspection and Maintenance Record). This write-up should include a description of the component temporarily added, removed, or relocated, and the statement "Change not entered on DD Form 365-3."

Q. How long can a temporary change in basic weight be carried forward on the DA Form 2408-13 or 2408-14?
A. For a period not to exceed 90 days.

Q. The temporary additions, removals, or relocations of aircraft equipment that are annotated on DA Form 2408-13 or 2408-14 must still be considered in weight and balance computations. How should these items be accounted for?
A. These changes should be treated as any other variation in loading. Use a weight and balance correction sheet to recompute weight and center-of-gravity.

Q. How often must a new weight & balance sheet be recomputed for aircraft?
A. This question was presented earlier under "AR 95-1." The answer provided was as stated in AR 95-1. AR 95-3, however, goes into a little more detail and that information is presented here.

All DD forms 365-4 in the aircraft weight and balance file as well as all duplicate forms (ie, in aircraft) must be checked for accuracy at least every 90 days and, if necessary, new forms prepared.

In addition, all weight and balance records will, as a minimum, be reviewed each year. This must include a weight and balance inventory of the aircraft and the following statement entered on DD Form 365-3: "Annual review and inventory completed." The date and the adjusted basic weight and moment must accompany this entry.

Q. If no changes are required during the 90 day weight and balance review, how is this annotated?
A. The DD Forms 365-4 should be redated and initialed in the date block to certify currency.

Q. When is an aircraft required to be re-weighed?
A. - Whenever an overhaul or major airframe repairs are accomplished.
- Whenever any modifications or component replacements (including painting) have been made for which the weight and center-of-gravity cannot be accurately computed.
- Whenever weight and center-of-gravity data records are suspected to be in error.
- Every 3 years for Class 1 aircraft; every 2 years for Class 2 aircraft.

PUBLICATION CYCLES

Q. How often is the GP published?
A. Every 32 weeks with a Planning Change Notice (PCN) issued at the 16-week midpoint of the publishing cycle. Urgent Change Notices (UCN) are issued as required.

Q. How often is the AP published?
A. AP/1, 2, 3, and 4 are published every 24 weeks with PCNs at 8- and 16-week intervals. UCNs are published as required. AP/1A, 2A, 3A, and 4A are published every 48 weeks with PCNs at 16- and 32-week intervals. The AP/1B is published every 8 weeks.

Q. What is the purpose of the Flight Information Handbook (FIH)?
A. This handbook contains aeronautical information required by DOD aircrews in flight, but which is not subject to frequent change.

Q. How often is the FIH published?
A. Every 32 weeks.

Q. What is the publication cycle for enroute low altitude charts (ELA), Area Charts, the IFR Supplement, and the terminal Low Altitude Instrument Approach Procedures (IAP) booklets?
A. 8 weeks.

Q. Which of the above publications has a Terminal Change Notice published at the 4-week midpoint?
A. The IAP booklets (12 volumes).

Q. What is the publication cycle for the VFR Supplement?
A. 24 weeks.

PUBLICATION CYCLE MATRIX

	4 wk	8 wk	16 wk	24 wk	32 wk	48 wk
GP			PCN		XXXXX	
AP/1		PCN	PCN	XXXXX		
AP/1A			PCN		PCN	XXXXX
AP/1B		XXXX				
FIH					XXXXX	
VFR SUPPL				XXXXX		
IFR SUPPL		XXXX				
ELA CHARTS		XXXX				
IAP	TCN	XXXX				
AREA CHARTS		XXXX				

PCN = Planning Change Notice
TCN = Terminal Change Notice

GENERAL PLANNING (GP)

Q. What are the two general categories of DOD FLIP planning documents?
A. General Planning (GP) and Area Planning (AP).

Q. What is the difference between General and Area Planning documents?
A. As can be ascertained by the name, the General Planning document contains general information on all FLIPs, explanation of terms (definitions), flight planning and codes, and aviation weather codes; Area Planning documents contain planning and procedure information for a specific region or geographic area.

NOTE: See Chapter 4 of this study guide for detailed questions concerning flight plans.

Q. IFR flight plans should be filed at least ______ minutes prior to the estimated time of departure.
A. 30 minutes (one hour in some areas).

Q. How should a flight plan be filed when departing a non-military airfield?
A. Either in person using FAA Form 7233-1, telephonically, or by aircraft radio

if no other means are available.

Q. After filing a flight plan in person or by telephone with Flight Service, what action is required upon departure?
A. Ensure that the actual departure time is passed to the FSS serving the departure area.

Q. What may be the result of not passing departure information to Flight Service?
A. You may arrive unannounced at your destination.

Q. An integrated NOTAM system exists between the FAA and the DOD. How can you tell if an enroute or destination facility is covered by this system?
A. A diamond symbol will be depicted between the installation name and the airfield identifier in the Enroute IFR Supplement.

Q. How do you check NOTAMS for an airfield or facility that is not covered by the FAA/DOD integrated NOTAM system?
A. Contact Flight Service.

Q. How do you contact Flight Service by telephone?
A. Dial 1-800-WX-BRIEF (992-7433).

Q. Civilian NOTAM coverage by the military NOTAM system is somewhat limited. For this reason, special steps should be taken regarding NOTAMS when a proposed flight will terminate at a civilian airport. Explain these special steps.
A. Aircrews should check FAA "D" and "L" series NOTAMS and the FAA "NOTICES TO AIRMEN" booklet.

Q. Explain how/where to look for "D" and "L" series NOTAMS.
A. "L" series NOTAMS are "Local" NOTAMS, which are similar to US Military AIRADS, and must be obtained from the FSS serving the destination airport (obtain the commercial phone number from the IFR Supplement or A/FD).

"D" series NOTAMS are "Distant" NOTAMS, which are similar to DOD NOTAMS, and may be obtained from any FSS.

Q. Where should you be able to find a copy of the "NOTICES TO AIRMEN"?
A. On the NOTAM Board in Base Ops or Flight Ops.

Q. Who is normally responsible for relaying IFR clearances to pilots of departing aircraft?
A. Many busy airports have a tower "clearance delivery" position established with a discrete radio frequency designated for this purpose. For airports having an operational control tower but no clearance delivery position, ATC IFR clearances are normally delivered by the tower "ground control" position.

NOTE: No visual surveillance or control over the movement of traffic is exercised by the tower "clearance delivery" position of operation.

Q. What items of information should be included in a pilots initial radio contact with the departure facility concerned (clearance delivery or ground)?

A. - Aircraft identification (ie, Cairns Clearance, Army 12345.....)
- Location on the airport (ie, ...
- Type of operation planned (ie, ...IFR....)
- First point of intended landing (ie, ...to
- Requested action (ie, ...clearance on request.)

Q. An IFR clearance to departing aircraft is issued prior to departure and should include what items of information, as appropriate (list in order)?

A. - Clearance prefix ("ATC CLEARS", "ATC ADVISES", "ATC REQUESTS", etc)
- Aircraft identification
- Clearance limit (destination, short range clearance, etc)
- Departure procedure or SID (headings and altitude, radar vectors, etc)
- Route of flight
- Altitude data in the order shown
- Holding instructions (for short range clearance limit, EFC time, etc)
- Any special information (clearance void time, hold for release, etc)
- Frequency and beacon information

Q. Occasionally you may receive an abbreviated IFR clearance which will not include all of the information of a detailed clearance. State two times when an abbreviated clearance should not be issued.

A. - If the route of flight originally filed with ATC has been changed by the pilot.
- When a detailed clearance is requested by the pilot.

Q. What is the requirement for reading back clearances?

A. There is no requirement for pilots to readback ATC clearances received while on the ground; however, pilots are responsible for clarifying any portion of the clearance that is not completely understood. In addition, controllers may request that pilots readback clearances.

Q. When communicating with a controlling agency, what items of information must be repeated for mutual verification?

A. Altimeter settings, vacating altitude for newly assigned altitude, change in heading for assigned vectors, and all "hold short" instructions, if applicable (ie, from tower or ground control).

Q. Initial clearance will include whenever practicable, the destination airport as the clearance limit. When this is not possible and you are cleared to a fix short of the filed destination you should either receive clearance beyond the fix or holding instructions at least how long prior to reaching the fix?

A. 5 minutes.

Q. If further clearance is not received (see above question) what should be your actions?

A. Contact ATC for instructions.

Q. When operating IFR in VFR conditions, whose responsibility is it to "see and

avoid" other aircraft?
A. When in radar contact you are only guaranteed separation from other IFR traffic and from participating VFR traffic. The pilot(s) is/are still responsible for traffic separation from other VFR traffic. For this reason, when flying IFR in VFR conditions one pilot should maintain constant vigilance for VFR aircraft (this is sometimes referred to as "using a safety pilot").

Q. You have received a combination clearance, IFR to VFR at a designated fix. You have been cleared IFR to the fix. If you wish to continue IFR beyond the clearance limit (that is, the fix), how far in advance should you contact ATC and inform them of your request?
A. At least 5 minutes prior to reaching the clearance limit.

Q. If the requested IFR clearance (see above question) is not received prior to reaching the clearance limit fix, what should you anticipate?
A. You should expect to hold at the fix until the clearance is received.

NOTE: See Chapter 8 of this study guide for detailed questions concerning "Holding".

Q. Explain the ATC clearance readback/altitude verification procedure for pilots of airborne aircraft.
A. Pilots are expected to read back those parts of ATC clearances which contain altitude assignments/restrictions or radar vectors as a means of mutual verification. The readback should be preceded by the aircraft identification and although it need not be verbatim, it should be made in the same sequence in which it was received.

Q. What is the proper radio procedure for verifying altitude clearances?
A. Include the newly assigned altitude when reporting vacating the previously assigned altitude (ie, "Army 12345, out of 3,000 for 2,500").

Q. Air Route Traffic Control Centers (ARTCC) are equipped to conduct direct communications with IFR traffic on VHF and UHF frequencies. IFR operations are expedited through the use of direct communications; however, to reduce frequency congestion, pilots are requested to use these frequencies strictly for communications pertinent to the control of IFR aircraft. Knowing this, what frequency should be used for miscellaneous communications such as filing flight plans or obtaining enroute weather?
A. Use the appropriate frequency for the nearest FSS or appropriate military facility.

Q. All aircraft are required to monitor what frequency for emergency purposes?
A. 243.0 MHz (Guard).

Q. What does ATC mean by a "squawk"?
A. This refers to the operation of your transponder.
"Stop squawk" - turn xponder OFF
"Squawk standby" - turn xponder to STBY

"Squawk 0250" - turn xponder to STBY, enter 0250, then turn back ON.

Q. When should the transponder be turned from the STBY to the ON position?
A. As late as practicable before takeoff (this is normally included as part of the "Before Takeoff" check).

Q. When should the transponder be turned from the ON to the STBY or OFF position?
A. As soon as practicable after landing (this is normally included as part of the "After Landing" check) or as otherwise directed by ATC.

Q. What transponder code is normally "squawked" for VFR operations?
A. 1200

Q. What transponder code should be "squawked" for emergencies?
A. 7700

Q. What transponder code should be "squawked" for lost commo?
A. 7600

Q. If you were to squawk "7500", what would ATC understand this to mean?
A. Your aircraft has been hijacked or forced to land to a new destination.

Q. If a code 7500 is squawked ATC may ask the pilot if the code is intentionally being used. What two ways can this be acknowledged?
A. By replying in the affirmative or by not replying at all.

Q. What are the five transponder modes of operation and what is each used for?
A. Mode 1 - military tactical; two numbers - 32 possible code combinations.
Mode 2 - military tactical; four numbers - 4096 possible code combinations.
Mode 3 - air traffic control (ATC); same as civilian mode A and therefore referred to as Mode 3/A. Four numbers - 4096 possible code combinations. This is the normal "squawk" code. Relays aircraft position to the controller.
Mode 4 - IFF (classified).
Mode C - relays the aircraft elevation (altitude) to the controller.

Q. ATC may sometimes request a speed adjustment in order to achieve or maintain required or desired spacing. What two times (as related to helicopters) should ATC **not** request a speed adjustment?
A. - While in a holding pattern; or
- While inside the FAF on final or a point 5 miles from the runway, whichever is closer.

Q. While being radar vectored by ATC, assigned IFR altitudes must be at or above the MVA. What is this and how is it determined?
A. Minimum Vectoring Altitude; this is determined by applying 1000′ above the highest obstacle in non-mountainous areas and 2000′ in mountainous areas.

Q. What is a "Low altitude alert"?

A. This may be used by ATC if a controller observes an aircraft to be at an altitude which, in the controller's best judgement, places the aircraft in an unsafe proximity to terrain or obstacles. The low altitude alert will be issued as follows: "Army 12345, low altitude alert, check your altitude immediately."

NOTE: The use of this alert is contingent on the capability of the controller to observe the unsafe condition.

Q. What is the proper procedure for closing a flight plan?
A. Military airfield - verbally confirm the closing of the flight plan with tower or Base Operations.
Non-military airfield - close the flight plan with Flight Service through any means of communication available.

Q. Are you authorized to call Flight Service long distance collect?
A. Yes, but you shouldn't have to with the 1-800-WX-BRIEF number. There are still a few areas left that do not have this option so a direct line must be used.

Q. What is "wake turbulence"?
A. It is a phenomena resulting from the passage of an aircraft through the atmosphere. The term includes vortices, thrust stream turbulence, jet wash, propeller wash, and rotor wash, both on the ground and in the air.

Q. What are the three basic effects of wake turbulence on a following aircraft?
A. - Imposed roll;
- Loss of altitude or rate of climb; and
- Possible structural stress.

Q. What is used to identify aircraft into classes for the application of wake turbulence criteria (separation procedures)?
A. Aircraft takeoff weight capacity.

Q. What are the three FAA classes of aircraft (see above question)?
A. Small - less than 12,500 lbs
Large - 12,500 lbs to 300,000 lbs
Heavy - 300,000 lbs or more

Q. Crew members of aircraft in the "heavy" class should include what word with their call signs during all communications?
A. "Heavy" (ie, "DELTA 562 HEAVY...").

Q. Describe wake turbulence avoidance procedures for the following situations:
(1) Landing behind a large aircraft on the same runway.
(2) Landing behind a large aircraft on a parallel runway.
(3) Landing behind a large aircraft on a crossing runway.
(4) Landing behind a departing large aircraft on the same runway.
(5) Landing behind a departing large aircraft on a crossing runway.
(6) Departing behind a large aircraft.

(7) Intersection takeoff on the same runway.
(8) Departing or landing after a large aircraft executing a low/missed approach, or touch-and-go landing, or whose final approach or takeoff flight path crossed over the runway of intended use.
(9) Enroute VFR flight.
A.
(1) Stay at or above the large aircraft's final approach flight path. Note the touchdown point of the larger aircraft and land beyond it.
(2) If the parallel runway is closer than 2500' consider possible drift to your runway. Stay at or above the large aircraft's final approach flight path. Note the touchdown point of the larger aircraft and land beyond it.
(3) Cross above the large aircraft's flight path.
(4) Note the large aircraft's rotation point and land well prior to it.
(5) Note the large aircraft's rotation point. If it is past the intersection continue the approach and land prior to the intersection. If the large aircraft rotates prior to the intersection avoid flight below the large aircraft's flight path. Abandon the approach unless a landing is assured well before reaching the intersection.
(6) Note large aircraft's rotation point. Make your rotation prior to this and continue climb above and upwind of the large aircraft's flight path until turning clear of the wake. Avoid subsequent headings which will cross below and behind a large aircraft.
(7) Be alert to large aircraft operations adjacent to and particularly upwind of your runway. If intersection takeoff clearance is received, avoid subsequent headings which will cross below a large aircraft's path.
(8) Because vortices settle and move laterally near the ground, the vortex hazard may exist along the runway and in your flight path after a large aircraft has completed the maneuver. You should assure that an interval of at least 2 minutes has elapsed before your takeoff or landing.
(9) Avoid flight below and behind a large aircraft's path. If a large aircraft is observed above on the same track (meeting or overtaking), adjust your position accordingly (preferably upwind).

NOTE: An acceptance from ATC of either of the following is an acknowledgement that you will ensure safe takeoff and landing intervals and that you accept the responsibility for providing your own wake turbulence separation:

- traffic information, or
- instructions to follow an aircraft, or
- the acceptance of a visual approach clearance.

AREA PLANNING (AP)

Q. What is the difference between the AP/1, the AP/1A, and the AP/1B?
A. The AP/1 contains planning and procedure information for North and South America. The AP/1A contains information regarding Special Use Airspace for this same area and the AP/1B contains information relative to Military Training Routes. The AP/1B also comes with six charts on three sheets containing graphic depictions of the IR, VR, and SR route system throughout the United States.

NOTE: See Chapter 11 of this study guide for questions related to airspace structure, to include Special Use Airspace.

Q. What is an MTR?
A. Military Training Route - airspace of defined vertical and lateral dimensions established for the conduct of military flight training at airspeeds in excess of 250 kias.

Q. MTRs will be preceded by the letters "IR" or "VR". What do these letters indicate?
A. IR - IFR Training Route; VR - VFR Training Route.

Q. MTRs are additionally identified by a three or four number character. What is the difference between these?
A. MTRs identified by three number characters have one or more segments above 1500' agl (ie, IR-206, VR-207); those MTRs with no segments above 1500' agl are identified by four number characters (ie, IR-1206, VR-1207).

FLIGHT INFORMATION HANDBOOK (FIH)

Q. Where in the FIH can you find information regarding emergency procedures?
A. Section A.

Q. When confronted with an emergency situation a "distress" or "urgency" call should be transmitted. What is the difference between these?
A. A "distress" is considered any situation where you are threatened by serious and/or imminent danger and require immediate assistance (ie, ditching, crash landing).

An "urgency" situation is one in which there is concern for the safety of an aircraft or of individuals on board but does not necessarily require immediate assistance.

Q. What words are transmitted as part of a distress call?
A. MAYDAY! MAYDAY! MAYDAY! (followed by aircraft identification and a description of the emergency).

Q. What words are transmitted as part of an urgency call?
A. PAN PAN! PAN PAN! PAN PAN! (followed by aircraft identification and a description of the emergency).

Q. What frequency should be used for an emergency call?
A. The ATC or flight following frequency in use at the time. If unable to establish communications on this frequency, utilize one of the emergency (Guard) frequencies.

Q. What are the VHF and UHF guard frequencies?
A. VHF - 121.5; UHF - 243.0.

Q. What actions should be taken with the transponder in the event of an

emergency?
A. Set the master control switch to the EMERGENCY position and proceed as indicated below:
- if under positive radar control maintain codes as previously set, or
- if not under positive radar control, set squawk to 7700.

Q. What elements of information, time permitting, should be transmitted with a distress or urgency call?
A.
- MAYDAY or PAN PAN call
- Name of station addressed.
- Aircraft identification and type.
- Nature of distress or urgency.
- Weather
- Pilot's intention (ditching, crash landing, etc) and request (fix, vectors, etc).
- Present position and heading.
- Altitude.
- Fuel remaining (in hours and minutes).
- Number of persons on board.
- Any additional information that might be helpful.

Q. How should an emergency call be canceled?
A. When no longer in distress immediately transmit a cancellation message on the same frequency or frequencies used for the distress message.

Q. What does SARSAT stand for?
A. Search and Rescue Satellite.

Q. How is SARSAT activated?
A. By keying the mike on a guard frequency (121.5 or 243.0) for 30 seconds or more.

Q. Because inadvertent activations of the SARSAT have been inordinately high, causing false alarms which seriously degrade the efficiency of the SAR system, it is recommended that transmissions on guard frequencies be limited to a ____ second duration except during actual emergency situations.
A. 15-second

Q. Two-way radio failure, also referred to as "lost commo", is the inability of an aircraft to communicate by radio. It is unlikely that all radios onboard an aircraft would fail at the same time so what may be initially perceived as a lost commo situation is, in most cases, something much simpler. Before initiating "lost commo" procedures it is advisable to utilize a common-sense approach to troubleshooting the problem. Describe such a procedure.
A. If you are unable to establish communications after a second or third attempt use the following steps (presented in order of precedence):

- Check to ensure that you are transmitting on the proper radio with the correct frequency "dialed-in". If your radio is set up for preset and/or manual frequency transmission ensure that the selector is in the proper position. If there is no "side tone" during transmission, ensure that the radio has not been inadvertently turned off. Check for presence of side tone on alternate frequency.

- Use an alternate transmission switch (ie, if you were using the cyclic transmit switch, try the floor switch).
- Have other pilot or crewmember attempt transmission (a problem could exist with your helmet mike or cyclic/floor transmission switches).
- If you have just changed frequencies IAW ATC instructions, return to the previous frequency and request an alternate new frequency.
- Use an alternate frequency or attempt communication on a different radio.
- If your aircraft is equipped with RETRANS capability, check to ensure that the selector lever is properly positioned.
- Attempt communications on a "guard" frequency.
- Make a call "in the blind" stating intentions.
- If possible, listen in on nearby navaids.

Q. As mentioned in the previous question, the likelihood of a complete loss of communications is remote. However, the FAA does have an emergency procedure for such an occurrence. If all troubleshooting fails and communications can not be re-established, how can you alert ATC that you are indeed "lost commo"?
A. The transponder should be adjusted to reply on Mode 3/A, Code 7600.

NOTE: Upon receiving this reply, pilots should expect ATC to attempt to communicate by transmitting on guard frequencies and available frequencies of navaids. It could be that one or more aircraft receivers are working but nothing is transmitting. In this case you may be requested by ATC to confirm receipt of a transmission by squawking "IDENT" or by some other means such as changing heading by 30° or more.

Q. If you are on an IFR flight plan at the time of radio failure but you are in VMC conditions, what should be your actions?
A. If able to maintain flight in VMC conditions continue flight under VFR and land as soon as practicable and notify ATC or Flight Service.

Several questions need to be answered if you are IFR in a lost commo situation (or if you are operating "on top" and unable to descend VMC prior to your destination) - the most important being (1) "How do I proceed from here to my destination?", (2) "What altitude do I use?", and (3) "What do I do when I reach my assigned clearance limit?" if there is one, or "When do I initiate my approach?" The next few questions deal with these issues.

Q. What are the ROUTE considerations when IFR (IMC) in a lost commo situation?
A. In order of precedence, proceed:
(1) By the <u>route assigned in the last ATC clearance</u>;
(2) If no route is assigned, proceed by the <u>route that ATC has advised may be expected in a further clearance</u>; or
(3) If neither of the above apply, by the <u>route filed in the flight plan</u>.

NOTE: If you are being radar vectored at the time of lost communications, proceed by the most direct route from the point of lost commo to the fix, route, or airway specified in the vector clearance and proceed as above.

Q. What ALTITUDE should be used if IFR (IMC) in a lost commo situation?
A. Use the highest of the following for the route segment being flown:
(1) The altitude assigned in the last ATC clearance;
(2) The altitude ATC has advised may be expected in a further clearance; or
(3) The minimum altitude for IFR operations (see Minimum IFR Altitude (MIA) in GP, "Explanation of Terms").

Q. You are IFR (IMC) in a lost commo situation. Your assigned clearance limit is a fix from which an approach begins (ie, IAF). How should you proceed upon reaching the fix?
A. You should have been issued an "Expect Further Clearance" (EFC) time in your clearance assignment. If you arrive at the fix prior to this time then you must enter holding and begin your descent or descent and approach as close as possible to the EFC time. If no EFC time was issued, begin the approach as close as possible to the expected time of arrival as filed on the flight plan (or amended with ATC).

Q. You are IFR (IMC) in a lost commo situation. Your assigned clearance limit is <u>not</u> a fix from which an approach begins. How should you proceed upon reaching the fix?
A. Again, in most cases an EFC time would have been issued. As in the case above you would enter holding at the fix if you arrive prior to the EFC time. Leave the assigned fix as close as possible to the EFC time and proceed to a fix from which an approach begins. Commence the descent or descent and approach as close as possible to the estimated time of arrival as filed on the flight plan (or amended with ATC). If no EFC time was issued, proceed through the assigned fix to a fix from which an approach begins and, as before, commence the approach as close as possible to the estimated time of arrival (may require holding at a fix from which an approach begins).

Q. When should lost communications procedures be initiated on a radar approach (ASR/PAR)?
A. If no transmissions are received from the ground controller for approximately one minute while being vectored to final, 15 seconds while on an ASR final approach, or 5 seconds while on a PAR final approach.

Q. What are the lost communications procedures for a radar approach?
A. (1) Attempt contact on a secondary frequency, the previously assigned frequency, the tower frequency, or guard.
(2) If unable to reestablish communications and unable to maintain VMC, proceed with a published instrument approach procedure or previously coordinated instructions. Change transponder to appropriate codes.
(3) Maintain the last assigned altitude or the minimum safe/sector altitude (emergency safe altitude if more than 25 NM from the facility), whichever is higher, until established on a segment of the published approach.

Q. Explain the following ATC light signals:
STEADY GREEN
FLASHING GREEN
STEADY RED
FLASHING RED
FLASHING WHITE
ALTERNATING RED & GREEN

A.

COLOR	ON GROUND	IN FLIGHT
STEADY GREEN	Cleared to T/O	Cleared to land
FLASHING GREEN	Cleared to taxi	Return for landing
STEADY RED	Stop	Give way to other acft and continue circling
FLASHING RED	Taxi clear of landing area (runway) in use	Airport unsafe-do not land
FLASHING WHITE	Return to starting point on airport	
ALT RED & GREEN	General warning signal - exercise extreme caution	

Q. All VFR operations (except VFR conditions on top) should utilize what primary frequency for position reporting?
A. 255.4 (FSS). If unable to establish communications on this frequency, utilize an alternate frequency for the nearest FSS or remoted outlet (check VFR sectional).

Q. Several transponder squawk codes have already been mentioned, namely 1200, 7500, 7600, and 7700. When might an aircraft be required to squawk 4000?
A. For aircraft authorized to operate in certain restricted/warning areas.

NOTE: A Mode 3/A Code 4000 may also be assigned by ATC to aircraft for operations requiring frequent or rapid changes in altitude.

Q. When are IFR position reports required to be made?
A. The answer to this question depends on whether or not you are in positive radar contact. If you are in radar contact then position reports are not required unless specifically requested by ATC.

When not in radar contact, position reports should be made at designated compulsory reporting points (indicated by a solid triangle on the ELA) or at non-compulsory reporting points (open triangle) when requested by ATC. Additionally, pilots should report over each reporting point used in the flight plan (to define the route of flight) for flights along a direct route (off airways).

Q. What phraseology should be used for establishing contact with a designated facility (initial radio contact in nonradar environment)?
A. - When a position report will be made: (Name) CENTER, (acft id), (position). (ie, "MEMPHIS CENTER, ARMY 12345, BOWLING GREEN")
- No position report: (Name) CENTER, (acft id), ESTIMATING (reporting point and time), AT (altitude), CLIMBING (or descending) TO MAINTAIN (altitude). (ie, "MEMPHIS CENTER, ARMY 12345, ESTIMATING BOWLING GREEN AT 35, 1,500 CLIMBING TO MAINTAIN 5,000").

Q. What phraseology should be used for establishing contact with a radar facility (such as approach or departure control)?
A. - On published route: (Name) APCH/DEP, (acft id), AT (altitude) CLIMBING TO MAINTAIN (or level at) (altitude).
(ie, "CAIRNS DEPARTURE, ARMY 12345, 1,200 CLIMBING TO MAINTAIN 5,000" or on a routine "handoff" to another controller, "MEMPHIS CENTER, ARMY 12345, LEVEL AT 5,000").
- On other than published route: same as above except pilot should include present navigational position (ie, "CAMPBELL APPROACH, ARMY 12345, 1,200 CLIMBING TO MAINTAIN 5,000, HEADING 070" or "DIRECT BOWLING GREEN").

Q. If you are in a radar environment and ATC advises "RADAR CONTACT LOST" or "RADAR SERVICE TERMINATED" you should resume normal position reporting. What are the elements of a complete position report?
A.
- Identification,
- Position,
- Time,
- Altitude,
- Type of flight plan (not required for IFR reports made directly to ARTCC or approach control),
- ETA and name of next reporting point,
- The name only of the next succeeding reporting point along the route of flight, and
- Pertinent remarks.

Q. In addition to position reports, when required, there are nine additional reports that must be made to ATC <u>at all times</u>. What are they?
A. Two reports have been combined on line (1) -
(1) Leaving an assigned altitude for a newly assigned altitude (including altitude changes while operating "VFR ON TOP"),
(2) Arriving at assigned holding (state time and altitude upon reaching assigned holding fix or point to which cleared),
(3) Leaving holding,
(4) Loss or impairment of navigational equipment (VOR, NDB, TACAN, ILS rcvr) while in controlled airspace,
(5) Missed approach, followed by next request (alternate airport, another approach, etc),
(6) Change in average TAS (at cruising altitude) of 5% or 10 kts (whichever is greater) from that filed in the flight plan,
(7) Unable to climb or descend at a rate of at least 500 fpm,

(8) Any information relating to safety of flight.

NOTE 1: In addition, all pilots encountering weather conditions which have not been forecast, or hazardous conditions which have been forecast, are expected to forward a report of such weather to ATC.

NOTE 2: The reports for reaching or leaving an assigned holding fix may be omitted by pilots of aircraft involved in instrument training at military terminal area facilities when radar service is being provided.

Q. In addition to the above reports, two other reports are required when not in "radar contact". What are they?
A.
(1) When leaving the final approach fix inbound on the final approach (nonprecision) or when leaving the outer marker or fix used in lieu of the outer marker inbound on the final approach (precision).
(2) A corrected estimate (ETA) at anytime it becomes apparent that an estimate as previously submitted is in error in excess of three minutes.

Q. What does PMSV stand for? How can you use it?
A. Pilot-to-Metro Service. This is basically a radio connection with a weather facility and the primary purpose is to communicate various types of weather information to pilots. Most stations are manned 24 hours and will have an observer, and in many cases, a forecaster. The radio call for PMSV is "METRO". In the case of Cairns, AAF, the PMSV frequency is 344.6. If you wished to use Cairns PMSV for weather information, or to update a DD Form 175-1, contact "Cairns METRO...." on UHF 344.6.

Q. Military pilots departing a location where military weather and NOTAM services are not available can get this information from one of two places. What are they?
A. - Nearest military facility having weather reporting capabilities (see list in Section C, FIH).
- An accredited local agency (FSS).

Q. Preflight weather briefings obtained from FSS or National Weather Service (NWS) briefer are available 24 hours a day in person, by radio, or by telephone. What are the three types of preflight briefings provided by FAA weather services?
A. - Standard briefing: request this briefing if you have not received recorded preliminary information such as a TWEB.
- Abbreviated briefing: request this briefing if you have received recorded preliminary information, need to update a previous briefing, or need only one or two specific items.
- Outlook briefing: request this briefing when your ETD is 6 or more hours from the briefing time. This is for planning purposes only and a Standard or Abbreviated briefing is needed prior to departure.

Q. What is a TWEB?
A. Transcribed Weather Broadcast - pre-recorded meteorological and aeronautical

data that is broadcast continuously on selected low frequency navigational aids. Such a navaid can be identified by a black circle with a white "T" (negative symbology) located in the upper right corner of the communication box. The information will be broadcast over the navaid frequency.

Q. What is HIWAS?
A. Hazardous Inflight Weather Advisory Service - a continuous broadcast of inflight weather advisories on VOR frequencies including summarized Severe Weather Forecast Alert, SIGMETs, Convective SIGMETs, Center Weather Advisories, AIRMETs, and PIREPs. VORs providing this service are identified by a small black box in the upper left corner of communications boxes on IFR charts and in the lower right corner of communication boxes on VFR charts.

Q. What is an EFAS? When is it available and how do you access it?
A. Enroute Flight Advisory Service - a service designed specifically to provide enroute aircraft with timely and meaningful weather advisories pertinent to the type of flight intended, route of flight, and altitude. This service is available throughout the conterminous U.S. from 6 am to 10 pm on a frequency of 122.0 MHz. Normal reception can be expected between 5,000′ agl and 17,500′ msl. EFAS is also known as "Flight Watch". Contact flight watch using the name of the parent facility, followed by your aircraft identification and append the name of the nearest VOR to your position. If you do not know what flight watch area you are flying in, initiate contact by using the words "FLIGHT WATCH", your identification, and the name of the nearest VOR.

Q. What is an AWOS?
A. An Automated Weather Observation System - a system consisting of various sensors, a processor, a computer-generated voice subsystem, and a transmitter to broadcast local minute-by-minute weather directly to the aircraft.

Q. What are the four operational levels of AWOS?
A.
- AWOS-A: reports only altimeter setting.
- AWOS-1: reports altimeter setting, wind data, temperature, dewpoint, and density altitude.
- AWOS-2: reports information in AWOS-1 plus visibility.
- AWOS-3: reports information in AWOS-2 plus cloud and ceiling data.

NOTE: AWOS information can normally be received within 25 NM of the AWOS site, at or above 3,000′ agl.

Q. An inflight weather report, also referred to as a Pilot Weather Report (PIREP) is <u>required</u> to be made under what conditions?
A.
- In-flight when requested.
- When unusual and unforecast weather conditions are encountered.
- When weather conditions on an IFR approach differ from the latest observation.
- When a missed approach is executed due to weather.
- When a wind shear is encountered on departure or arrival.

Q. What is the format for a PIREP?
A. - Location of phenomena
- Time (UTC)
- Altitude (msl)
- Type aircraft
- Skycover (bases, tops, and amount)
- Flight Visibility and weather
- Air temperature
- Wind
- Turbulence
- Icing
- Remarks

Q. Turbulence can be reported as either LIGHT, MODERATE, SEVERE, or EXTREME. Explain the difference between these.
A. - LIGHT: turbulence that momentarily causes slight, erratic changes in attitude and/or altitude.
- MODERATE TURBULENCE: turbulence that cause changes in altitude and/or attitude, but with the aircraft remaining in positive control at all times. It usually causes variations in indicated airspeed.
-- MODERATE CHOP: turbulence that causes rapid bumps or jolts without appreciable changes in aircraft altitude or attitude.
- SEVERE: turbulence that causes large, abrupt changes in altitude and/or attitude. It usually causes large variations in indicated airspeed. Aircraft may be momentarily out of control.
- EXTREME: turbulence in which the aircraft is violently tossed about and is practically impossible to control. It may cause structural damage.

Q. At what altitude should the altimeter setting automatically be changed to 29.92?
A. 18,000' msl

Q. NOTAM code groups contain a total of five letters, the first of which is always the letter _____.
A. "Q"

Q. What is the purpose of the additional letters in a NOTAM code group?
A. The second and third letters identify the subject reported upon. The fourth and fifth letters denote its status of operation.

Q. The NOTAM code is used in conjunction with the NOTAM Class I format. The subject matter of the NOTAM will be subdivided into items identified by the letters A through G, each followed by a closing parenthesis. What is the content of each item?
A.
A) Location identifier of the facility reported on
B) Date/time group consisting of eight figures, indicating the beginning of the period of validity giving month, day, and hour in UTC or with immediate effect (WIE).

C) Approximate duration (APRX DUR) of the period of validity, or permanent (PERM), or until further notice (UFN), or date/time group giving month, day, and hour in UTC.
D) Specified periods (if applicable) of activity of hazards, status, or condition being reported on, within the period of validity.
E) Text of the NOTAM, coded, amplified, or completed as necessary. Plain language will be used when suitable NOTAM code groups are not available.
F) Lower limit of navigation warnings/airspace reservations if applicable.
G) Upper limit applicable to Item F).

NOTE: If there is no information for entry against a particular item, that item/identifier will be omitted.

Q. What is the difference between a Class I and a Class II NOTAM?
A. Class I NOTAMS are reported by facilities and updated on a continual basis. The Class I NOTAM format consists of up to seven lines of information (Items A through G) containing NOTAM code groups. A NOTAM code group contains a total of five letters, the first of which is always the letter "Q".
Class II NOTAMS are more permanent in nature and/or reflect changes to currently published FLIPs. These NOTAMS are provided in a book titled "NOTICES TO AIRMEN," which is published every 14 days. Class II NOTAMS are not provided during pilot briefings (with FSS) unless specifically requested.

ENROUTE IFR SUPPLEMENT (ES)

Q. The IFR Supplement contains an Airport/Facility Directory (A/FD). What are the selection criteria for airports listed in the Supplement?
A. Must have at least one of the following:
- A published DOD instrument approach procedure and/or ASR/PAR RADAR approach minima.
- Selected airports/heliports specifically requested by a U.S. military service.

NOTE: Some airports located in Canada and Mexico will be included if portrayed on enroute charts for emergency use.

Q. In addition to airports/heliports, what other information (in alphabetical order) is presented in the Supplement?
A.
- Air Route Traffic Control Centers (ARTCCs)
- Flight Information Centers/Flight Service Stations (FIC/FSS)
- Communications Stations
- Radio Aids to Navigation

Q. In the Enroute Supplement all bearings, radials, courses, and tracks are given in degrees of __________ (magnetic/true/grid) north.
A. Magnetic

Q. All mileages are given as __________ (statute/nautical) miles; all times are ___________ (local/zulu); and all elevations are in feet above or below ________

(the ground/mean sea level).
A. NM; UTC (zulu); msl.

Q. Immediately following the airport/facility name may sometimes be another name in parentheses. What does this mean?
A. The name in parentheses is an alternate name.

Q. What is indicated by the open diamond symbol immediately following the names of some facilities?
A. This symbol indicates that the airfield is covered by the FAA/DOD integrated NOTAM system.

NOTE: Aircrews filing to airports not covered by this system should contact a FSS for appropriate NOTAMS.

Q. The location identifier is an assigned code having _____ characters.
A. 4 characters for ICAO codes, 3 characters for FAA assigned codes.

Q. Immediately following the location identifier is a code for the operating agency. What code is used for U.S. Army facilities? What code is used to denote civil airports wherein a permit covers use by transient military aircraft?
A. "A"; "P".

Q. All facilities in the A/FD will include the geographic location in degrees, minutes, and tenths of a minute. Immediately following the location entry some airfields will have another entry in parentheses. What is this?
A. The UTM grid coordinates (if available).

Q. Following the facility location will be the airport elevation, given in feet above or below mean sea level (msl). What point on the airfield is used as a reference for this figure?
A. The highest point on the landing surface.

Q. When a facility elevation is sea level it will be indicated as "00". How are elevations below sea level indicated?
A. By a minus (-) sign preceding the figure.

Q. Next in the A/FD format comes the time conversion information. Explain how this is used to obtain local standard time.
A. The information provided gives you the number of hours to be added to or subtracted from UTC in order to obtain local standard time. For example, an entry of "UTC-4" indicates that you should subtract 4 hours from UTC to get local time. In areas where daylight savings time is observed, the conversion is included in parentheses (ie, "UTC+4(+5DT)" would tell you to add 4 hours to UTC to obtain local time except when daylight savings time was in effect, then you would add 5 hours).

Q. All facilities in the Enroute Supplement operate continuously unless otherwise specified. What does the symbol "++" following facility operating hours

indicate?
A. This means that, during periods of daylight saving time, the effective hours of operation will be one hour earlier than shown.

Q. When does daylight saving time go into effect and how long does it remain in effect?
A. From 0200 local time on the first Sunday in April to 0200 local time on the last Sunday in October.

Q. All of the conterminous States observe daylight saving time except whom?
A. Arizona and that portion of Indiana lying in the Eastern standard time zone.

Q. What does the symbol "(B)" in the A/FD format indicate?
A. A rotating beacon is operational from sunset to sunrise. An omission of this symbol indicates that a rotating beacon is not available or does not operate from sunset to sunrise.

Q. Airport runways are normally numbered in relation to their magnetic direction rounded off to the nearest _____°.
A. 10

Q. Runway dimensions are expressed in ________ (feet/yards/meters).
A. Feet

Q. Runway surface material is either classified as "Hard" or "Other". Two of the most common hard surface compositions are designated "ASP" and "CON". What type of surface does each of these represent?
A. ASP - asphalt; CON - concrete.

Q. The RUNWAY BEARING STRENGTH of an airfield represents that runway's weight bearing capacity. How is this figure expressed?
A. In thousands of pounds (add 000 to the figure).

Q. Instead of providing a weight bearing capacity, some airfields provide what is known as a "PCN". What is this?
A. Pavement Classification Number - a five-part code consisting of an ACN followed by runway data concerning type of pavement, pavement subgrade category, maximum authorized tire pressure, and pavement evaluation method.

Q. Where would you expect to find information regarding the type of fuel available at a particular airport?
A. Under the section marked SERVICE.

Q. What is a JASU?
A. Jet Aircraft Starting Unit.

Q. What is meant by a fuel entry that is preceded by the letters "NC" and enclosed in parentheses?
A. The fuel (listed in the parentheses with the NC) is noncontract. This means

that the refueling agent does not have a valid refueling contract with the government and should be used only as a last resort. Plan refueling stops accordingly.

NOTE: Where contract fuel is available the name of the refueling agency will be provided in parentheses.

Q. What is used to purchase contract fuel?
A. DD Form 1896, US Military Aircraft Identaplate (Jet Fuel).

Q. In the event that noncontract fuel must be purchased, what should be used to pay for it?
A. SF 149, US Government National Credit Card.

Q. If the REMARKS section of an airfield mentions that a PPR is required what does this mean?
A. PPR, or Prior Permission Required, means just that - you must contact the airfield manager or representative and request permission to fly there (or get fuel, etc). Usually a PPR "number" will be issued. Hold on to this along with the name or initials of the person issuing the PPR. Don't forget to also include any PPR information on your flight plan in the REMARKS section.

Q. At what other times might a PPR be required even though not specifically stated?
A. If the words "OFFL BUS ONLY" (official business only) appear somewhere in the REMARKS section or if transient alert services (fuel, etc) are required outside the published duty hours. In addition, all aircraft carrying hazardous materials must obtain prior permission as outlined in appropriate regulations.

Q. Do the remarks "OFFL BUS ONLY" or "PPR" for an airfield preclude the use of that airfield as an alternate for IFR flight?
A. No.

Q. Explain the meaning of the following COMMUNICATIONS line:
APP CON - (R) (E) 123.6 117.7T 122.1R 122.4x
A. This facility's approach control has (R)adar capability. The (E) means that it monitors VHF <u>and</u> UHF guard frequencies. You can contact approach control on 123.6 or 122.1. Be advised that 122.1 is receive only and they can transmit but not receive on 117.7. 122.4 is available on request.

Q. If the facility has a (U) or (V) instead on an (E) what does this mean? What if there is neither an (E), (U), or (V)?
A. A (U) indicates that the facility only monitors the UHF guard frequency; a (V) the VHF guard frequency. If neither of these are indicated then the facility does not monitor a guard frequency.

Q. Any radio aids to navigation listed under NAVAIDS will be those having the same name as the facility, or those having different names but are located at the airfield or used to provide final approach course guidance to that airfield. In

order for a navaid to be considered "at the field" it must not be more than how far from the airport landing surface?
A. 1 NM. Will also specify "At Fld".

Q. Frequency protection, where applicable, will be indicated immediately following the navaid identifier. What is meant by the following frequency protection information: VOR - (H) 113.9 AGS (100/50)?
A. Normal interference-free reception can be anticipated out to 100 NM from the VOR and up to 50,000′ msl.

Q. Given the following navaid information:
NAVAIDS - CLARKSVILLE VOR-DME - (T) 110.6 CKV CH43 36°37.3′N 87°27.8′W At Fld. 540/1°00′W SNUFF NDB - L 335 CK 36°31.6′N 87°23.2′W 347° 5.9 NM to Fld. Unk/1°00′W VOR unuse 220°-310° byd 10 NM blw 6000′ (this entry has been modified for instructional purposes). Explain it.
A. The Clarksville VORTAC has a normal anticipated interference-free service out to 25 NM and up to 12,000′. The frequency is 110.6 and the identifier is CKV. DME information can be obtained by tuning receiver to channel 43. Its location (in Lat/Long) is given as well as a description. In this case, the VORTAC is located within 1 NM of the airport landing surface. The antenna elevation is 540′ msl and the magnetic variation is 1° West.

Snuff NDB has a compass locator (less than 25 watts, 15 NM) with an ADF frequency of 336. Its identifier is CK and its lat/long location is provided. Additionally we can see that the antenna is located 5.9 NM from the field on a magnetic heading of 347° (from the NDB). The height of the antenna is reported as "unknown" and the magnetic variation is 1° West.

Finally, there is mention that the VOR is unusable (reliable reception cannot be assured) between 220° and 310° from the navaid when below 6000′ msl and outside of 10 NM.

Q. Most frequencies will be shown in light type. What is indicated by a frequency that is shown in **bold** type?
A. That frequency is for high altitude use only.

Q. What is indicated by a frequency that is preceded by a small "d"?
A. This is a discrete frequency.

Q. What is an ADIZ?
A. Air Defense Identification Zone - an airspace of defined dimensions (see AP/1) within which certain rules for the security of aircraft are mandatory in the interest of National Security.

Q. What type of flight plan is required to be filed when penetrating any part of an ADIZ?
A. Defense Visual Flight Rule (DVFR) or IFR.

Q. What are the allowable tolerances for adherence to an ADIZ flight plan?
A. - Time: +/- 5 minutes of estimated penetration
- Distance: 10 NM from course centerline (over land)

20 NM from course centerline (over water)
- Altitude: none

Q. What is a CTAF?
A. Common Traffic Advisory Frequency. This frequency may be a UNICOM, MULTICOM, FSS, or tower frequency and is identified in the appropriate publications.

Q. What are the recommended procedures for the use of CTAF?
A. All inbound traffic should monitor and communicate as appropriate on the designated CTAF from 10 miles to landing. Departure aircraft should monitor/communicate on the appropriate frequency from start-up, during taxi, and until 10 miles from the airport unless otherwise specified in FARs or local procedures.

If an airport has a tower and it is temporarily closed, or operated on a part-time basis and there is no FSS on the airport or the FSS is closed, use the CTAF (usually the tower local control frequency) to self-announce your position or intentions. If there is a UNICOM station in operation on the airport, the wind direction and runway in use should be obtained from the UNICOM station. The wind direction and runway information may not be available on UNICOM frequency 122.950. Where there is no tower, FSS, or UNICOM station on the airport, use MULTICOM frequency 122.9 for self-announce procedures. Such airports will be identified in appropriate publications.

SUMMARY OF RECOMMENDED COMMUNICATIONS PROCEDURES

(see next page)

SUMMARY OF RECOMMENDED COMMUNICATIONS PROCEDURES

FACILITY AT AIRPORT	FREQUENCY USE	OUTBOUND	INBOUND
UNICOM (No Tower or FSS)	Communicate with UNICOM station on published CTAF frequency (122.7, 122.8, 122.725, 122.975, or 123.0), If unable to contact UNICOM station, use self-announce proced on CTAF.	Before taxiing and before taxiing on the runway for dep	10 miles out. Entering downwind, base, and final. Leaving the runway.
No Tower, FSS or UNICOM	Self-announce on MULTICOM frequency 122.9	Same as above	Same as above
No Tower in operation, FSS open.	Communicate with FSS on CTAF.	Same as above	Same as above
FSS closed (No Tower)	Self-announce on CTAF....	Same as above	Same as above
Tower or FSS not in operation	Self-announce on CTAF....	Same as above	Same as above

Q. The IFR Supplement can also be used to aid in giving position reports, changing a flight plan, or filing a flight plan in flight. Where can you find this information "neatly" arranged for you?
A. On the back cover (inside back cover for FAA procedures).

AIRSPACE STRUCTURE, ATC FACILITIES and IFR COMMUNICATIONS

AIRSPACE STRUCTURE

Q. What DOD FLIP provides the dimensions and operating requirements for controlled airspace within the United States?
A. AP/1

Q. On 16 SEP 93 all FAA Controlled Airspace was reclassified to conform with ICAO Annex 11. There are now six classes of airspace, designated by the letters A through E and G. What types of airspace did each formerly represent?
A. A - Positive Control Areas (PCAs)
B - Terminal Control Areas (TCAs)
C - Airport Radar Service Areas (ARSAs)
D - Control Zones (CZ)/Airport Traffic Areas (ATAs)
E - Control Zone/nontower airports and controlled airspace, including Victor airways, transition areas, and control areas
G - Uncontrolled airspace

Q. What is the difference between "controlled" and "uncontrolled" airspace?
A. Controlled airspace is that airspace designated as either Class A, Class B, Class C, Class D, or Class E. It has defined dimensions within which air traffic control service is provided to IFR flights and to VFR flights IAW the airspace classification.

Uncontrolled airspace is that airspace <u>not</u> designated as either Class A, Class B, Class C, Class D, or Class E. Uncontrolled airspace is designated as Class G airspace and is an area within which ATC does not have the authority or responsibility to exercise control over air traffic.

Q. Describe the dimensions of Class A airspace.
A. Generally, that airspace from 18,000' msl up to and including FL600, including the airspace overlying the waters within 12 NM of the coast of the 48 contiguous states and Alaska.

Q. What are the basic VFR weather minimums for operations within Class A airspace?
A. VFR minimums are not applicable to Class A airspace (see question below).

Q. Unless otherwise authorized, no aircraft may be operated in Class A airspace unless on what type of flight plan?
A. IFR

Q. What are the dimensions of Class B airspace?
A. <u>As depicted</u>. Generally, Class B airspace begins at the surface and extends upward to 10,000' msl surrounding the nation's busiest airports. The configuration of each Class B airspace area is individually tailored and consists of a surface area and two or more layers (resembling an upside-down wedding cake), and is designed to contain all published instrument procedures once an aircraft enters the airspace.

Q. In order to operate VFR in Class B airspace what equipment requirements must

be met?
A. - Must have two-way radio communications; and
- Must have an operable radar beacon transponder with automatic altitude reporting equipment (operable Modes 3/A, C), unless otherwise authorized by ATC.

Q. All Class B airspace has a "Mode C veil". Explain what this is.
A. The Mode C veil is depicted graphically on IFR charts and is an area, generally having a 30 NM radius around the primary airport on which the airspace is located, inside which all aircraft must be equipped with a transponder having Mode C capability (unless otherwise approved by ATC).

Q. At what altitude does the Mode C veil terminate?
A. 10,000′ msl

Q. Is a clearance required for flights within the 30-NM veil if operations remain outside of Class B airspace (beneath the "layers" of the wedding cake)?
A. No.

Q. ATC may, upon notification, immediately authorize a deviation from the altitude reporting requirement (Mode C); however, a request for a deviation from the 4096 transponder equipment requirement (Mode 3/A) must be submitted to the controlling ATC at least how far in advance of the proposed operation?
A. 1 hour.

Q. In addition to the above VFR requirements, what additional equipment is required for IFR flights within Class B airspace?
A. An operable VOR or TACAN receiver.

Q. Equipment requirements for operation within Class B airspace have already been mentioned. What communication requirements are there for operating within this airspace?
A. Must have an ATC clearance prior to entering.

Q. Your initial call to a Class B controlling ATC receives the response, "Army 12345, stand by." Are you cleared to enter the airspace?
A. No, merely establishing two-way communications does not constitute a clearance to enter Class B airspace. You <u>must</u> hear something like, "Army 12345, radar contact, <u>cleared to enter</u> Atlanta Class B airspace, proceed as ......"

Q. Aircraft also require a clearance to depart Class B airspace (ground departure). How is this obtained?
A. From Clearance Delivery - advise them of your intended altitude and route of flight. This applies to VFR departures as well as IFR departures.

Q. Aircraft not landing or departing the primary airport may obtain an ATC clearance to transit the Class B airspace when traffic conditions permit. Such VFR aircraft are encouraged, to the extent possible, to operate at altitudes above or below the Class B airspace or transit through established VFR corridors. Pilots operating in VFR corridors are urged to exchange aircraft position

information on what VHF frequency?
A. 122.750 MHz.

Q. A VFR corridor defines airspace of specified vertical and lateral boundaries through Class B airspace. It is surrounded on all sides by Class B airspace. Is an ATC clearance or communications with ATC required when using a VFR corridor?
A. No.

Q. What is the maximum indicated airspeed authorized for all aircraft operated below 10,000′ msl?
A. 250 kts (288 mph), unless authorized by the Administrator (ie, Military Training Routes (MTRs)).

Q. What is the maximum indicated airspeed authorized for aircraft operating within Class B airspace?
A. 250 kts (288 mph).

Q. What is the maximum indicated airspeed authorized for aircraft operating in airspace underlying Class B airspace or in VFR corridors?
A. 200 kts (230 mph)

Q. What are the visibility requirements for VFR operations within Class B airspace?
A. 3 miles.

Q. What are the cloud clearance requirements for VFR operations within Class B airspace?
A. Clear of clouds.

Q. What are the dimensions of Class C airspace?
A. As depicted. Class C airspace has a basic design with minor site-specific variations. The design consists of two concentric circles both centered on the primary airport.

The inner circle has a radius of 5 NM and the outer circle has a radius of 10 NM. The airspace of the inner circle extends from the surface of the primary airport to 4,000′ agl. The airspace of the area between 5 and 10 NM begins at 1,200′ agl and extends to the same altitude cap as the inner circle.

In addition, an uncharted outer area, normally having a radius of 20 NM, extends from the lower limits of radar/radio coverage up to the ceiling of the approach control's delegated airspace.

Q. What equipment is required for operations within Class C airspace?
A. - Must have two-way radio communications; and
- Must have an operable radar beacon transponder with automatic altitude reporting equipment (operable Modes 3/A, C), unless otherwise authorized by ATC.

NOTE 1: Transponder requirements also exist for overflight of Class C airspace up to 10,000′ msl.

NOTE 2: All aircraft operating above 10,000' msl are required to have an operable transponder.

Q. In addition to equipment, what other requirements must be met prior to operating within Class C airspace?
A. Must establish two-way communications with the ATC facility providing ATC services prior to entry and thereafter maintain those communications while in Class C airspace.

Q. What information should be included in your initial contact with ATC?
A. Position, altitude, transponder code, destination, and request for Class C service.

NOTE: Radio contact should be initiated far enough from the Class C airspace boundary to preclude entering Class C airspace before two-way radio communications are established.

Q. Your initial call to a Class C controlling ATC receives the response, "Aircraft calling Nashville Approach, stand by." Have you met the criteria for establishing two-way communications?
A. No, ATC must respond with the aircraft call sign. If you heard something like, "Army 12345, stand by", then radio communications have been established and, unless otherwise advised by ATC, you can enter Class C airspace.

NOTE: Depending on work load or traffic condition, you may be instructed by ATC to remain outside the airspace until conditions permit entry (ie, "Army 12345, remain outside Class C airspace and standby").

Q. What are the communications procedures for departing a primary or satellite airport with an operating control tower (within Class C airspace)?
A. Two-way radio communications must be established and maintained with the control tower, and thereafter as instructed by ATC while operating in Class C airspace.

Q. What are the communications procedures for departing a satellite airport <u>without</u> an operating control tower (within Class C airspace)?
A. Two-way radio communications must be established with the ATC having jurisdiction over the Class C airspace as soon as practicable after departure.

Q. What is the maximum indicated airspeed authorized for flights at or below 2,500' agl and within 4 NM of the primary airport of a Class C airspace area?
A. 200 kts (230 mph).

Q. What are the visibility requirements for VFR operations within Class C airspace?
A. 3 miles.

Q. What are the cloud clearance requirements for VFR operations within Class C airspace?

A. Basic Cloud Clearance - 500' below, 1,000' above, 2,000' horizontally.

Q. What are the cloud clearance requirements for SVFR operations within Class C airspace?
A. Clear of clouds (ATC authorization required for entry into or departure from Class C airspace under Special VFR conditions).

Q. What services are provided by ATC within Class C airspace?
A. - Sequencing of all arriving aircraft to the primary airport.
- Standard IFR separation between IFR aircraft.
- Traffic advisories and conflict resolution between IFR and VFR aircraft.

Q. What services are provided by ATC within the outer area of Class C airspace?
A. Same services as within the Class C airspace when two-way radio communication and radar contact is established.

Q. Is pilot participation within the outer area of Class C airspace mandatory?
A. No. Aircraft participation in this area is strongly encouraged but not a VFR requirement. Radar service is automatic unless discontinued at pilot request.

Q. What are the dimensions of Class D airspace?
A. As depicted. Generally, that airspace from the surface to 2,500' above the airport elevation (charted in MSL) surrounding those airports that have an operational control tower. The configuration of each Class D airspace area is individually tailored but is normally a circular area of 5 SM radius, including any extensions necessary for instrument departure and arrival paths. A specific Class D airspace is based on one primary airport with an operational control tower, but may include other airports.

Q. Any extensions that are part of the basic surface area are charted with blue segmented lines. When the extension is outside of Class D airspace it is indicated by magenta segmented lines. What does this mean?
A. The area outlined by magenta segmented lines indicates a separate surface area and represents Class E airspace.

Q. What is the general rule of thumb for determining whether or not extensions for instrument approaches will be part of the Class D surface area or an extension?
A. If all extensions are 2 miles or less, they remain part of the Class D surface area. However, if any one extension is greater than 2 miles, then all extensions become Class E.

Q. What equipment requirements are there for operations within Class D airspace?
A. Unless otherwise authorized by ATC, an operable two-way radio is required.

Q. What are the VFR arrival/through-flight requirements for Class D airspace?
A. Unless landing to or departing from an airport within the Class D airspace all aircraft must avoid the airspace (unless otherwise authorized by ATC). In addition, two-way communications must be maintained with the control tower while

operating within the Class D airspace, to include ground movement areas.

Q. What are the communications procedures for departing a primary or satellite airport (within Class D airspace) with an operating control tower?
A. Two-way radio communications must be established and maintained with the control tower, and thereafter as instructed by ATC while operating in the Class D airspace.

Q. What are the communications procedures for departing a satellite airport (within Class D airspace) without an operating control tower?
A. Two-way radio communications must be established with the ATC facility having jurisdiction over the Class D airspace as soon as practicable after departure.

Q. Pilots of arriving aircraft must contact the control tower on the publicized frequency. What information should be provided on initial contact?
A. Position, altitude, destination, and any request(s).

NOTE: Radio contact should be initiated far enough in advance of the Class D airspace boundary to preclude entering Class D airspace before two-way radio communications are established.

Q. What is the maximum indicated airspeed authorized for operations within Class D airspace?
A. 200 kts (230 mph).

Q. Ceiling height for Class D airspace is charted in feet MSL. How is this depicted on sectional charts?
A. As two numbers contained in a box of blue segmented lines (brackets).

Q. How is Class D airspace identified on an ELA chart?
A. By the boxed-in letter "D".

Q. At those airports where the control tower does not operate 24 hours a day, the operating hours of the tower will be listed on the appropriate charts and in the A/FD. What happens to the airspace classification when the control tower is not in operation?
A. The Class D airspace reverts to a surface-based Class E airspace, and Class E surface area rules are applicable.

Q. How is a part-time tower depicted on an ELA chart?
A. With a star next to the boxed-in letter "D".

Q. What are the visibility requirements for VFR operations within Class D airspace?
A. 3 miles

Q. What are the cloud clearance requirements for VFR operations within Class D airspace?
A. Basic Cloud Clearance = 500′ below, 1000′ above, 2000′ horizontally.

Q. What are the minimum ceiling and visibility requirements for VFR flight within Class C, Class D, or surface-based Class E airspace?
A. 1000' ceiling (agl) and 3 miles visibility.

Q. When ceiling and visibility for the above-mentioned airspaces drops below 1,000 - 3 the airspace is considered IFR and VFR aircraft must obtain a SVFR clearance for arrival or departure. What are the cloud clearance requirements for SVFR operations within Class C, Class D, or surface-based Class E airspace?
A. Clear of clouds.

Q. What is the general definition of Class E airspace?
A. If airspace is controlled airspace but is not classified as either Class A, Class B, Class C, or Class D, then it is Class E airspace.

Q. What are the vertical limits of Class E airspace?
A. Airspace extending upward from either the surface or a designated altitude to overlying or adjacent controlled airspace.

Q. What are the equipment requirements for operating within Class E airspace?
A. None

Q. What are the communication requirements for entry into or through-flight of Class E airspace?
A. No specific requirements.

Q. What are the five types of Class E airspace?
A.
- Surface area
- Extension
- Transition area
- Enroute area
- Federal airway

Q. How is a Class E surface area depicted and what is its purpose?
A. It is depicted by segmented magenta lines on a sectional chart and is configured to contain all instrument approach procedures for an airfield.

Q. What is the purpose of a Class E extension area?
A. These are airspace extensions for Class B, Class C, or Class D surface areas which provide for standard instrument approach procedures without communications requirements for VFR operations.

Q. How is Class E airspace for transition areas depicted on the sectional, what is its purpose and at what altitude does it begin?
A. The magenta shaded areas surrounding airfields with instrument approaches allow IFR traffic to remain within controlled airspace while transitioning to or from the terminal or enroute environment. Class E airspace around such airfields begins at 700' agl within the shaded area and 1200' agl outside of the shaded area (unless it abuts Class G airspace, in which case you will also see blue shading).

Q. All federal airways are considered Class E airspace. What are the dimensions of such airspace?
A. Normally, 4 NM either side of centerline of a victor airway and extending from 1200' agl up to but not including 18,000' msl.

Q. What are the visibility requirements for VFR operations within Class E airspace?
A. Less than 10,000' MSL............ 3 miles
At or above 10,000' MSL.......... 5 miles

Q. What are the cloud clearance requirements for VFR operations within Class E airspace?
A. Less than 10,000' MSL............ Basic Cloud Clearance
At or above 10,000' MSL.......... 1,000' below
1,000' above
1 statute mile horizontally

Q. What is Class F airspace?
A. This is an ICAO airspace classification wherein ATC provides separation services to IFR aircraft when practical - it does not have an equivalent U.S. classification.

Q. What is the general definition of Class G airspace?
A. That portion of the airspace that has not been designated as Class A, Class B, Class C, Class D, or Class E airspace - also referred to as "uncontrolled" airspace.

Q. What are the visibility requirements for VFR operations within Class G airspace when at or below 1,200' agl (regardless of MSL altitude)?
A. Day = ½ mile; Night = 1 mile

Q. What are the cloud clearance requirements for VFR operations within Class G airspace when at or below 1,200' agl (regardless of MSL altitude)?
A. Clear of clouds.

NOTE: The above visibility and cloud clearance requirements are for rotary wing aircraft only and are based on exceptions to fixed-wing criteria provided by FAR 91.155 and Table 5-1, AR 95-1.

Q. What are the visibility and cloud clearance requirements for VFR operations within Class G airspace when above 1,200' agl?
A. Above 1,200' agl but less than 10,000' msl:

Day............. 1 mile; Basic Cloud Clearance
Night........... 3 miles; Basic Cloud Clearance

Above 1,200' agl and above 10,000' msl:

Day/Night........ 5 miles;

1,000′ below, 1,000′ above, 1 mile horiz

Q. What are the IFR altitude requirements for operations within Class G airspace (certain areas of the U.S.)?
A. Must maintain Minimum IFR Altitude (MIA) or above at all times. MIA represents an altitude with at least 1,000′ of obstacle clearance (2,000′ in designated mountainous areas) within a horizontal distance of 4 NM from the course to be flown. In addition, there are altitude requirements based on the semi-circular rule (see next question).

Q. Discuss the semi-circular rule for both IFR and VFR operations.
A. IFR: use an odd thousand altitude if your heading is between 0° and 179° (magnetic); use an even thousand altitude if your heading is between 180° and 359° (magnetic).

VFR: for VFR flights above 3,000′ agl, use an odd thousand altitude plus 500′ if your heading is between 0° and 179° (magnetic); use an even thousand altitude plus 500′ if your heading is between 180° and 359° (magnetic).

MEMORY AID: NEODD - north thru east = odd thousand (plus 500′ for VFR)
SWEVEN - south thru west = even thousand (plus 500′ for VFR)

Q. What is meant by the "Hierarchy of Airspace Classes"?
A. In the event that there is an overlap of airspace classes, Class A preempts Class B, Class B preempts Class C, and so on. When overlapping airspace designations apply to the same airspace, the operating rules associated with the more restrictive airspace will be used.

Q. In addition to the different classes of airspace, there exists airspace of defined dimensions where activities must be confined because of their nature and/or where limitations may be imposed on aircraft operations that are not part of those activities. What is this airspace called?
A. Special Use Airspace (SUA).

Q. There are six types of special use airspace. Name them.
A. - Prohibited Area
- Restricted Area
- Warning Area
- Military Operations Area (MOA)
- Alert Area
- Controlled Firing Areas (CFA)

Q. Of the six types of special use airspace, which are considered regulatory?
A. Prohibited Areas and Restricted Areas (the remaining four types are considered nonregulatory).

Q. Special use airspace are charted on IFR and visual charts and include the hours of operation, altitudes, and the controlling agency. What is the one exception to this and why?
A. Controlled Firing Areas are not charted since they do not cause a

nonparticipating aircraft to change its flight path. Activities within a CFA are suspended immediately when spotter aircraft, radar, or ground lookout positions indicate an aircraft might be approaching the area.

Q. Describe a Prohibited Area.
A. Airspace of defined dimensions within which the flight of aircraft is prohibited.

Q. What is the primary reason for the establishment of prohibited areas?
A. Such areas are established for security or other reasons associated with the national welfare.

Q. Where are prohibited areas currently located?
A. Amarillo, TX; Kennebunkport, Maine; Mount Vernon, VA; Thurmont, Maryland; and Washington, D.C.

Q. Describe a Restricted Area.
A. Airspace of defined dimensions within which the flight of aircraft, while not wholly prohibited, is subject to restrictions. Restricted areas denote the existence of unusual, often invisible hazards to aircraft such as artillery firing, aerial gunnery, or guided missiles.

Q. What is the procedure for VFR aircraft to penetrate or transit a restricted area?
A. Approval must be obtained by the pilot from the using or controlling agency prior to penetration or transit. If approval is not obtained the area must be circumnavigated.

NOTE: Penetration of a restricted area without authorization from the using or controlling agency may be extremely hazardous to an aircraft and its occupants.

Q. What is the procedure for IFR aircraft to penetrate or transit a restricted area?
A. ATC is responsible for either obtaining clearance for IFR aircraft to transit a restricted area or for vectoring the aircraft around the restricted area (this includes aircraft cleared to maintain VFR-ON-TOP).

Q. Describe a Warning Area.
A. Airspace which may contain hazards to nonparticipating aircraft in international airspace. Warning Areas are established beyond the 3-mile limit along U.S. coastlines. Though the activities conducted within Warning Areas may be as hazardous as those in Restricted Areas, Warning Areas cannot be legally designated as Restricted Areas because they lie over international waters.

NOTE: Penetration of a Warning Area during periods of activity may be hazardous to an aircraft and its occupants.

Q. Describe a Military Operations Area (MOA).
A. Airspace of defined vertical and lateral limits established for the purpose

of separating certain military training activities from IFR traffic. Most training activities within a MOA necessitate acrobatic or abrupt flight maneuvers.

Q. Pilots operating under VFR should exercise extreme caution while flying within an MOA when military activity is being conducted. How can you determine the activity status of an MOA?
A. Contact any FSS within 100 NM of the area. Also, contact the controlling agency for advisories prior to entering an active MOA.

Q. Describe an Alert Area.
A. Airspace of defined dimensions used to inform nonparticipating pilots of areas that may contain a high volume of pilot training or an unusual type of aerial activity.

Q. Are there any restrictions or requirements for penetrating or transiting an Alert Area?
A. No, but pilots should be particularly alert when flying in these areas.

Q. Certain aspects of a Controlled Firing Area have already been discussed. For what reason does a CFA exist?
A. To contain activities which, if not conducted in a controlled environment, could be hazardous to nonparticipating aircraft.

Q. Other airspace areas include Airport Traffic Advisory Areas and Military Training Routes (MTRs). Describe these areas of airspace.
A. Airport Advisory Area - the area within 10 NM of an airport where a control tower is not operating but where a FSS is located. At such locations, the FSS provides advisory service to arriving and departing aircraft.

Military Training Route - routes developed for use by the military for the purpose of conducting low-altitude, high-speed training. Generally, MTRs are established below 10,000′ msl for operations at speeds in excess of 250 kts. The routes above 1,500′ agl are developed to be flown, to the maximum extent possible, under IFR. The routes at 1,500′ agl and below are generally developed to be flown under VFR.

Q. Are pilots required to participate in the Local Airport Advisory program?
A. No, participation is not mandatory but strongly recommended.

Q. Where would you expect to find detailed information regarding MTRs?
A. In the Area Planning (AP/1B).

ATC FACILITIES and CLEARANCES

Q. What is "Air Traffic Control" (ATC)?
A. This is a service provided by an appropriate authority to promote the safe, orderly, and expeditious flow of air traffic.

Q. FAR 91.115 states that no person may operate an aircraft in controlled

airspace in IFR unless what two conditions are met?
A. - A flight plan is filed; and
- An ATC clearance is received.

Q. What is meant by an ATC "clearance"?
A. This is an authorization by ATC for known air traffic to proceed under specific traffic conditions within controlled airspace. The primary purpose of a clearance is to prevent collisions between aircraft.

Q. No pilot in command may deviate from an ATC clearance unless an amended clearance has been obtained from ATC. What are some exceptions to this?
A. - In an emergency
- After canceling IFR to VFR
- When an ATC clearance will cause deviations from rules and regulations (in this case you must so advise ATC and get amended instructions).

Q. When ATC issues a clearance or instructions, pilots are expected to take specific actions to comply with ATC instructions upon receipt of the instructions. What "specific actions" are expected upon assignment of heading and/or altitude changes by ATC?
A. Heading assignment - immediately comply with new heading assignment and acknowledge with ATC (ie, "ARMY 12345, right to 350°").

Altitude assignment - immediately begin complying with new altitude assignment and acknowledge with ATC by giving vacating altitude followed by "climbing to maintain" or "descending to maintain" and new altitude (ie, "ARMY 12345 is out of 1,500, climbing to maintain 3,000"). Advise ATC if unable to climb/descend at a rate of at least 500 fpm.

Q. ATC phraseology must be understood in order for the pilot to comply with ATC's wishes. Explain the following phraseology.
"IMMEDIATELY..."
"AT PILOT'S DISCRETION...."
"CROSS (FIX) AT, AT OR ABOVE, AT OR BELOW (ALTITUDE)...."
A. The term "immediately" is used by ATC when such action is required to avoid an imminent situation; "at pilot's discretion" is used to indicate that the pilot has the option of when to initiate a specific action within the parameters specified; and "cross at (above)(below)..." is used to specify an altitude restriction at a designated fix.

Q. What is the purpose of a clearance delivery facility?
A. To relay IFR clearances to departing aircraft.

Q. Where is clearance delivery normally located?
A. At towered airports having a high volume of air traffic a separate position is designated for clearance delivery and is given a discrete frequency. At other towered airports clearance delivery is relayed through the ground control position. At nontowered airports, clearances are delivered through flight service.

Q. Given the following agencies, what is the order of priority for obtaining an IFR clearance: approach control, tower, clearance delivery, ground control, FSS?

A. 1. Clearance Delivery
   2. Ground Control
   3. Tower
   4. FSS
   5. Approach Control

Q. What are three ways that an IFR clearance can be obtained from Flight Service?

A. In person, by phone, or by radio.

Q. An IFR clearance to departing aircraft is issued prior to departure and should include what items of information, as appropriate (list in order)?

A. - Clearance prefix (if applicable)
   - Aircraft identification
   - Clearance limit
   - Departure procedure or SID
   - Route of flight
   - Altitude data in the order shown
   - Holding instructions
   - Any special information
   - Frequency and beacon information

Q. What information would be included in the clearance prefix?

A. "ATC CLEARS", "ATC ADVISES", "ATC REQUESTS", etc.

Q. Explain the meanings of the above clearance prefixes.

A. "ATC CLEARS" - used to prefix an ATC clearance when it is relayed to an aircraft by other than an air traffic controller (ie, flight service).

"ATC ADVISES" - used to prefix a message of noncontrol information when it is relayed by other than an air traffic controller.

"ATC REQUESTS" - used to prefix an ATC request when it is relayed to an aircraft by other than an air traffic controller.

Q. Occasionally you may receive an abbreviated IFR clearance which will not include all of the information of a detailed clearance. State two times when an abbreviated clearance should not be issued.

A. - If the route of flight originally filed with ATC has been changed by the pilot.
   - When a detailed clearance is requested by the pilot.

Q. What is the requirement for reading back clearances?

A. There is no requirement for pilots to readback ATC clearances received while on the ground; however, pilots are responsible for clarifying any portion of the clearance that is not completely understood. In addition, controllers may request that pilots readback clearances.

Q. When communicating with a controlling agency, what items of information must be repeated for mutual verification?

A. Altimeter settings, vacating altitude for newly assigned altitude, change in heading for assigned vectors, and all "hold short" instructions, if applicable (ie, from tower or ground control).

Q. A clearance will include, whenever practicable, the destination airport as the clearance limit. However, ATC may utilize a short range clearance procedure in lieu of a clearance to the destination airport. If this occurs, when should you expect to receive additional clearance?
A. At least 5 minutes prior to reaching the clearance fix.

Q. If further clearance is not received (see above question) what should be your actions?
A. Contact ATC for instructions.

Q. When operating IFR in VFR conditions, whose responsibility is it to "see and avoid" other aircraft?
A. When in radar contact you are only guaranteed separation from other IFR traffic and from participating VFR traffic. The pilot(s) is/are still responsible for traffic separation from other VFR traffic. For this reason, when flying IFR in VFR conditions one pilot should maintain constant vigilance for VFR aircraft (this is sometimes referred to as "using a safety pilot").

Q. You have received a combination clearance, IFR to VFR at a designated fix. You have been cleared IFR to the fix. If you wish to continue IFR beyond the clearance limit (that is, the fix), how far in advance should you contact ATC and inform them of your request?
A. At least 5 minutes prior to reaching the clearance limit.

Q. If the requested IFR clearance (see above question) is not received prior to reaching the clearance limit fix, what should you anticipate?
A. You should expect to hold at the fix until the clearance is received.

NOTE: See Chapter 8 of this study guide for detailed questions concerning "Holding".

Q. Explain the ATC clearance readback/altitude verification procedure for pilots of airborne aircraft.
A. Pilots are expected to read back those parts of ATC clearances which contain altitude assignments/restrictions or radar vectors as a means of mutual verification. The readback should be preceded by the aircraft identification and although it need not be verbatim, it should be made in the same sequence in which it was received.

Q. What is the proper radio procedure for verifying altitude clearances?
A. Include the newly assigned altitude when reporting vacating the previously assigned altitude (ie, "Army 12345, out of 3,000 for 2,500").

Q. What information might be included with the departure procedure or SID line in the clearance format?

A. SIDs or SID cancellation, headings and altitude, radar vectors.

Q. When would holding instructions be issued in the clearance format?
A. If ATC is using a short range clearance procedure in lieu of clearance to destination. Holding instructions are normally issued at this time but must be issued at least 5 minutes prior to reaching the clearance fix.

Q. In addition to holding instructions, what other information should be issued at this time?
A. An Expect Further Clearance (EFC) time.

Q. What items of information might be included in the "special instructions" line of the clearance format?
A. A clearance void time, hold for release instructions, etc.

Q. What requirements are there for reading back clearances?
A. No requirement exists for pilots to readback ATC clearances received while on the ground; however, pilots should clarify any portion of a clearance that is not completely understood. In addition, controllers may request that pilots readback any clearances.

Q. What is the purpose of Ground Control?
A. When a Ground Control position exists at a towered facility, its purpose is to direct the movement of aircraft on the airport surface. In addition, Ground Control relays ATC clearances when no clearance delivery position exists.

Q. What frequencies are normally reserved for Ground Control?
A. 121.6 - 121.9 MHz

NOTE: Controllers may omit the number preceding the decimal point. For instance, if a controller instructs you to "Contact ground point 9", you should automatically know that the frequency is 121.9.

Q. Selected terminal areas have a continuous broadcast of recorded noncontrol information known as ATIS. What does this stand for and what is its primary purpose?
A. Automatic Terminal Information Service; purpose is to improve controller effectiveness and to relieve frequency congestion by automating the repetitive transmission of essential but routine information.

Q. What information would you expect to find in an ATIS broadcast?
A. Information on ceilings, visibility, obstructions to visibility, temperature, dew point, wind direction, wind speed, altimeter setting, remarks, and instrument approach(es) and runway(s) in use.

Q. When an ATIS broadcast does not contain information on ceiling, sky condition, or visibility, what does this mean?
A. This indicates that a ceiling or cloud layers are at or above 5,000 feet and visibility of 5 SM or better.

Q. An ATIS broadcast says, "Advise on initial contact you have information Golf." What does this mean?
A. On initial contact with Ground Control or Tower let them know that you have the latest ATIS information (in this case, "information Golf") so that they do not have to repeat it when issuing hover/taxi/departure instructions (ie, CAIRNS GROUND, ARMY 12345, PARKED A-17 FOR HOVER TO CHARLIE PAD, IFR TO LAWSON, WITH INFORMATION GOLF").

Q. A control tower is a terminal facility that uses air/ground radio and visual signaling to provide ATC to aircraft operating in the vicinity of an airport. This includes authorizing aircraft to taxi (through ground control when available), takeoff, and land at the airport. What items of information should be included in a clearance for arriving aircraft?
A. - Specific traffic pattern information
- Runway in use
- Wind
- Altimeter setting
- Supplementary information
- Clearance to land

NOTE: The task of repeating some of the above information can be relieved from the controller by listening to the appropriate ATIS frequency for relevant information and so advising the tower on initial contact that you "have information (whatever)."

Q. What does the phrase "With the Numbers" or "Have Numbers" mean when communicating with tower?
A. This indicates that the pilot has received wind, runway, and altimeter information ONLY (ie, recently overheard takeoff clearance for another aircraft), and the tower does not have to repeat this information. THIS DOES NOT INDICATE RECEIPT OF AN ATIS BROADCAST AND SHOULD NEVER BE USED FOR THIS PURPOSE.

Q. What items of information should be included in a clearance for departing aircraft?
A. - Wind
- Runway takeoff clearance
- Special instructions

Q. What is the purpose of Departure Control?
A. This is a function of the approach control facility and provides information to departing IFR aircraft or, on occasion, enroute transitioning aircraft, and under certain conditions, to VFR aircraft. This is the first facility to establish radar contact (if radar is available) and issues special instructions or amendments to clearances in the terminal area during departure.

Q. Approach Control is a <u>terminal</u> ATC facility that uses radar and nonradar capabilities to provide approach control services to aircraft arriving, departing, or transiting through the airspace controlled by the facility, at one or more airports within its area. A U.S. Army Approach Control facility is most

often referred to as what?
A. ARAC (Army Radar Approach Control).

Q. When receiving terminal information through Approach Control, what information should you expect to receive?
A. - "EXPECT TYPE APPROACH ________"
- "RUNWAY IN USE ________"
- "WINDS ________"
- "CEILING AND VISIBILITY ________"
- "ALTIMETER ________"

Q. What information should you expect to receive through Approach Control for an approach clearance?
A. - "CLEARED FOR ________ APPROACH"
- "CIRCLE TO RUNWAY ________" (if applicable)
- "CONTACT TOWER ON ________ (freq)" (or advisory frequency)
- "LOCATION ________"

Q. If Approach Control is a terminal ATC facility, what facility provides enroute ATC services?
A. An Air Route Traffic Control Center, or ARTCC (pronounced "ART' see"). There are 24 ARTCCs which are divided into sectors. Each sector has its own discrete frequency which is handled by either a single controller or a team of controllers. As a flight progresses from one sector to another the pilot is directed to change to the appropriate sector frequency.

Q. A sector controller can monitor the progress of flights within their sectors even though they may be hundreds of miles from the aircraft being monitored. What allows a sector controller to do this?
A. A network of Air Route Surveillance Radar (ARSR). There are over 100 ARSRs in the U.S., each one typically covering 200 NM up to 40,000 feet. Several radars are required for each ARTCC with radar scopes normally centrally located at the ARTCC facility through the use of remotes.

Q. List some of the services that are provided by a Flight Service Station.
A. - Pilot briefings
- Enroute communications
- VFR search and rescue
- Assist lost aircraft and aircraft in distress
- Relay ATC communications
- Originate notices to airmen (NOTAMs)
- Broadcasting aviation weather
- Receive and process IFR flight plans
- Monitoring NAVAIDS

Q. What is the purpose of the aircraft transponder?
A. This enables the aircraft to identify itself automatically when properly challenged by friendly surface and airborne radar equipment.

Q. "Transponder" is short for what?
A. Transmitter-Responder.

Q. The transponder operates in conjunction with _________ radar and is classified as ___________ radar.
A. Primary; secondary.

Q. What are the five transponder modes of operation and what is each used for?
A. Mode 1 - military tactical; two numbers - 32 possible code combinations.
Mode 2 - military tactical; four numbers - 4096 possible code combinations.
Mode 3 - air traffic control (ATC); same as civilian mode A and therefore referred to as Mode 3/A. Four numbers - 4096 possible code combinations. This is the normal "squawk" code. Relays aircraft position to the controller.
Mode 4 - IFF (classified).
Mode C - relays the aircraft elevation (altitude) to the controller.

Q. What is the warm-up time required for the transponder?
A. 2 minutes.

Q. What are the five selector modes on the master control switch of the transponder?
A. OFF - turns set off.
STBY - places set in warm-up.
LOW - reduces receiver sensitivity.
NORM - normal receiver sensitivity.
EMER - transmits emergency reply signals to Modes 1, 2, or 3/A interrogations regardless of mode control settings.

NOTE: Newer versions of the transponder have only four selector modes - OFF, STBY, NORM, and EMER.

Q. What does ATC mean by a "squawk"?
A. This refers to the operation of your transponder.
"Stop squawk" - turn xponder OFF
"Squawk standby" - turn xponder to STBY
"Squawk 0250" - turn xponder to STBY, enter 0250, then turn back ON.

Q. ATC tells you to "stop altitude squawk." What is meant by this?
A. If the Mode C portion of your transponder is malfunctioning and reporting an altitude which is more than 300' off your actual altitude, you may be requested by ATC to "stop altitude squawk," or in other words - turn off Mode C.

Q. When should the transponder be turned from the STBY to the ON position?
A. As late as practicable before takeoff (this is normally included as part of the "Before Takeoff" check).

Q. When should the transponder be turned from the ON to the STBY or OFF position?
A. As soon as practicable after landing (this is normally included as part of the

"After Landing" check) or as otherwise directed by ATC.

Q. Mode 3/A has 4096 possible codes used for air traffic control. These codes are given numerical values from 0001 to 7700. What code is normally "squawked" for VFR operations?
A. 1200

Q. What transponder code should be "squawked" for emergencies?
A. 7700

Q. What transponder code should be "squawked" for lost commo?
A. 7600

Q. If you were to squawk "7500", what would ATC understand this to mean?
A. Your aircraft has been hijacked or forced to land to a new destination.

Q. If a code 7500 is squawked ATC may ask the pilot if the code is intentionally being used. What two ways can this be acknowledged?
A. By replying in the affirmative or by not replying at all.

IFR COMMUNICATIONS

Q. What items of information should be included in a pilot's initial radio contact with clearance delivery?
A. - Aircraft identification (ie, Cairns Clearance, Army 12345.....)
- Type of operation planned (ie, ...IFR.....)
- First point of intended landing (ie, ...to Lawson.....)
- Requested action (ie, ...request clearance)

Q. When are IFR position reports required to be made?
A. The answer to this question depends on whether or not you are in positive radar contact. If you are in radar contact then position reports are not required unless specifically requested by ATC.

When not in radar contact, position reports should be made at designated compulsory reporting points (indicated by a solid triangle on the ELA) or at non-compulsory reporting points (open triangle) when requested by ATC. Additionally, pilots should report over each reporting point used in the flight plan (to define the route of flight) for flights along a direct route (off airways).

Q. When necessary to be made, what are the elements of a complete position report?
A. - Identification,
- Position,
- Time,
- Altitude,
- Type of flight plan (not required for IFR reports made directly to ARTCC or approach control),
- ETA and name of next reporting point,
- The name only of the next succeeding reporting point along the route of

flight, and
- Pertinent remarks.

Q. In addition to position reports, when required, there are nine additional reports that must be made to ATC at all times. What are they?
A. Two reports have been combined on line (1) -
(1) Leaving an assigned altitude for a newly assigned altitude (including altitude changes while operating "VFR ON TOP"),
(2) Arriving at assigned holding (state time and altitude upon reaching assigned holding fix or point to which cleared),
(3) Leaving holding,
(4) Loss or impairment of navigational equipment (VOR, NDB, TACAN, ILS rcvr) while in controlled airspace,
(5) Missed approach, followed by next request (alternate airport, another approach, etc),
(6) Change in average TAS (at cruising altitude) of 5% or 10 kts (whichever is greater) from that filed in the flight plan,
(7) Unable to climb or descend at a rate of at least 500 fpm,
(8) Any information relating to safety of flight.

NOTE 1: In addition, all pilots encountering weather conditions which have not been forecast, or hazardous conditions which have been forecast, are expected to forward a report of such weather to ATC.

NOTE 2: The reports for reaching or leaving an assigned holding fix may be omitted by pilots of aircraft involved in instrument training at military terminal area facilities when radar service is being provided.

Q. In addition to all of the above reports, two other reports are required when not in "radar contact". What are they?
A.
(1) When leaving the final approach fix inbound on the final approach (nonprecision) or when leaving the outer marker or fix used in lieu of the outer marker inbound on the final approach (precision).
(2) A corrected estimate (ETA) at anytime it becomes apparent that an estimate as previously submitted is in error in excess of three minutes.

Q. What phraseology should be used for establishing contact with a radar facility (such as approach or departure control)?
A. - On published route: (Name) APCH/DEP, (acft id), AT (altitude) CLIMBING TO MAINTAIN (or level at) (altitude).
(ie, "CAIRNS DEPARTURE, ARMY 12345, 1,200 CLIMBING TO MAINTAIN 5,000" or on a routine "handoff" to another controller, "MEMPHIS CENTER, ARMY 12345, LEVEL AT 5,000").
- On other than published route: same as above except pilot should include present navigational position (ie, "CAMPBELL APPROACH, ARMY 12345, 1,200 CLIMBING TO MAINTAIN 5,000, HEADING 070" or "DIRECT BOWLING GREEN").

Q. What phraseology should be used for establishing contact with a designated

facility (initial radio contact in nonradar environment)?
A. - When a position report will be made: (Name) CENTER, (acft id), (position). (ie, "MEMPHIS CENTER, ARMY 12345, BOWLING GREEN")
- No position report: (Name) CENTER, (acft id), ESTIMATING (reporting point and time), AT (altitude), CLIMBING (or descending) TO MAINTAIN (altitude). (ie, "MEMPHIS CENTER, ARMY 12345, ESTIMATING BOWLING GREEN AT 35, 1,500 CLIMBING TO MAINTAIN 5,000").

IFR COMMUNICATIONS MATRIX

(see next page)

IFR COMMUNICATIONS MATRIX

RADAR ENVIRONMENT	
INITIAL CONTACT	* ID * ALT (level, C/D to maintain) ** HEADING ** DIRECT (navaid) * only if on other than published airway
POSITION REPORT	None Required, but if requested give: * ID * POSITION

NON-RADAR ENVIRONMENT	
INITIAL CONTACT (compulsory reporting point)	* ID * POSITION (followed by full position report)
INITIAL CONTACT (at other than compulsory reporting points)	* ID * ETA/NAME OF NEXT REPORTING POINT * ALT (level, C/D to maintain)
POSITION REPORT	* ID * POSITION * TIME * ALTITUDE * ETA/NAME OF NEXT REPORTING POINT * NAME ONLY OF NEXT SUCCEEDING REPORTING POINT * REMARKS

Q. The likelihood of a complete loss of communications is remote. However, the FAA does have an emergency procedure for such an occurrence. If all troubleshooting fails and communications can not be re-established, how can you alert ATC that you are indeed "lost commo"?

A. The transponder should be adjusted to reply on Mode 3/A, Code 7600.

NOTE: Upon receiving this reply, pilots should expect ATC to attempt to communicate by transmitting on guard frequencies and available frequencies of navaids. It could be that one or more aircraft receivers are working but nothing is transmitting. In this case you may be requested by ATC to confirm receipt of a transmission by squawking "IDENT" or by some other means such as changing heading by 30° or more.

Q. If you are on an IFR flight plan at the time of radio failure but you are in VMC conditions, what should be your actions?
A. If able to maintain flight in VMC conditions continue flight under VFR and land as soon as practicable and notify ATC or Flight Service.

Several questions need to be answered if you are IFR in a lost commo situation (or if you are operating "on top" and unable to descend VMC prior to your destination) - the most important being (1) "How do I proceed from here to my destination?", (2) "What altitude do I use?", and (3) "What do I do when I reach my assigned clearance limit?" if there is one, or "When do I initiate my approach?" The next few questions deal with these issues.

Q. What are the ROUTE considerations when IFR (IMC) in a lost commo situation?
A. In order of precedence, proceed:
(1) By the route assigned in the last ATC clearance;
(2) If no route is assigned, proceed by the route that ATC has advised may be expected in a further clearance; or
(3) If neither of the above apply, by the route filed in the flight plan.

NOTE: If you are being radar vectored at the time of lost communications, proceed by the most direct route from the point of lost commo to the fix, route, or airway specified in the vector clearance and proceed as above.

Q. What ALTITUDE should be used if IFR (IMC) in a lost commo situation?
A. Use the highest of the following for the route segment being flown:
(1) The altitude assigned in the last ATC clearance;
(2) The altitude ATC has advised may be expected in a further clearance; or
(3) The minimum altitude for IFR operations (see Minimum IFR Altitude (MIA) in GP, "Explanation of Terms").

Q. You are IFR (IMC) in a lost commo situation. Your assigned clearance limit is a fix from which an approach begins (ie, IAF). How should you proceed upon reaching the fix?
A. You should have been issued an "Expect Further Clearance" (EFC) time in your clearance assignment. If you arrive at the fix prior to this time then you must enter holding and begin your descent or descent and approach as close as possible to the EFC time. If no EFC time was issued, begin the approach as close as possible to the expected time of arrival as filed on the flight plan (or amended with ATC).

Q. You are IFR (IMC) in a lost commo situation. Your assigned clearance limit is not a fix from which an approach begins. How should you proceed upon reaching

the fix?
A. Again, in most cases an EFC time would have been issued. As in the case above you would enter holding at the fix if you arrive prior to the EFC time. Leave the assigned fix as close as possible to the EFC time and proceed to a fix from which an approach begins. Commence the descent or descent and approach as close as possible to the estimated time of arrival as filed on the flight plan (or amended with ATC). If no EFC time was issued, proceed through the assigned fix to a fix from which an approach begins and, as before, commence the approach as close as possible to the estimated time of arrival (may require holding at a fix from which an approach begins).

Q. When should lost communications procedures be initiated on a radar approach (ASR/PAR)?
A. If no transmissions are received from the ground controller for approximately one minute while being vectored to final, 15 seconds while on an ASR final approach, or 5 seconds while on a PAR final approach.

Q. What are the lost communications procedures for a radar approach?
A. (1) Attempt contact on a secondary frequency, the previously assigned frequency, the tower frequency, or guard.
(2) If unable to reestablish communications and unable to maintain VMC, proceed with a published instrument approach procedure or previously coordinated instructions. Change transponder to appropriate codes.
(3) Maintain the last assigned altitude or the minimum safe/sector altitude (emergency safe altitude if more than 25 NM from the facility), whichever is higher, until established on a segment of the published approach.

IFR KNEEBOARD INFORMATION

AVIONICS CHECK
- VOR/MARKER BEACON
- ADF
- TRANSPONDER
- DOPPLER/GPS
- RADAR ALTIMETER

ATIS INFORMATION [ ]
Ceiling ____________________
Visibility ____________________
Temperature ____________________
Dew Point ____________________
Wind Direction/Speed ____________________
Altimeter Setting ____________________
Instrument Approach(es) ____________________
Runway in Use ____________________
Remarks ____________________

AIRSPACE STRUCTURE, ATC FACILITIES and IFR COMMUNICATIONS

CLEARANCE
Aircraft ID ________________________________
Clearance Limit ________________________________
Departure Procedure/SID ________________________________
Route ________________________________
Altitude Data ________________________________
Special Information ________________________________
Frequency and Squawk ________________________________

DEPARTURE INFORMATION
Runway in Use ________________________________
Surface Winds ________________________________
Altimeter Setting ________________________________
Time (when requested) ________________________________
Ceiling/Visibility (below VFR) ________________________________

HOVER/TAXI INSTRUCTIONS
Hover ________________________________
Winds ________________________________
Altimeter ________________________________

HOLDING INSTRUCTIONS
Hold (direction from airway) ________________________________
Name of Holding Fix ________________________________
Holding Radial, Bearing, or Airway ________________________________
Type Holding Pattern [ ] Left Turns [ ] Right Turns
Expect Further Clearance Time ________________________________

APPROACH BRIEF
Type Approach ________________________________
Runway ________________________________
Location ________________________________
MDA/DH ________________________________
Time ________________________________
Missed Approach Procedure ________________________________

TERMINAL INFORMATION
Expect Type Approach ________________________________
RWY in Use ________________________________
Winds ________________________________
Ceiling/Visibility ________________________________
Altimeter Setting ________________________________

APPROACH CLEARANCE
Cleared for ________________ Approach
(circle-to-land RWY ________)
Altitude ________________________________
Contact Tower/Advisory (freq) ________________________________
Location ________________________________

IFR FLIGHT PLANNING

Q. What does AR 95-1 state about preflight planning?
A. "The aviator will evaluate aircraft performance, approach and departure data, notices to airmen (NOTAMs), and the Army Aviation Flight Information Bulletin."

Q. What are some places where a weather briefing may be obtained?
A. Military or civilian forecaster - in person, by telephone, or by radio (last resort).

Q. A military weather briefing should be completed on what form?
A. DD Form 175-1 (Flight Weather Briefing), often referred to as simply a "-1".

Q. As a minimum, a weather briefing should include what items of information?
A. Enroute, destination, and alternate (if used) weather information. More specifically:

- A forecast for destination and alternate airfields at ETA to include:
  - -- ceiling and visibility
  - -- weather phenomena producing low ceilings and visibility
  - -- hazards to flight, such as thunderstorms, icing, gusty winds, and high density altitude
  - -- height of cloud tops
- An enroute forecast to destination and alternate airfields, to include:
  - -- hazards to flight
  - -- freezing level
  - -- height of cloud tops and bases
  - -- flight level winds and temperatures
- An overall weather picture, including locations of frontal systems, high- and low-pressure areas, the rate and direction of their movement, and the associated weather conditions.

Q. How long is a weather forecast good for?
A. 1 Hour and 30 minutes from the time the forecast is received provided the aircraft to which it applies has not departed.

Q. Can a weather forecast (see above question) be extended?
A. Yes, by the pilot-in-command after coordination with a weather facility.

Q. Where should you look to determine the minimum equipment required for VFR/IFR flight?
A. AR 95-1, Table 5-2.

Q. When should you file for **direct** routes of flight?
A. Only if the mission requires it or if considerable savings of fuel or time can be realized.

Q. What are the two basic considerations when planning a direct route?
A.
- FAR 91.119 must be complied with to ensure obstruction clearance; and
- NAVAID ranges and restrictions must be researched to ensure reception for the entire leg.

IFR FLIGHT PLANNING

Q. Why should you NOT plan a flight at MEA if the flight level temperature will be significantly below standard?
A. Because the lowering of pressure levels in air significantly colder than standard will result in the true altitude being significantly lower than the indicated altitude.

Q. How is ETE determined for an IFR flight?
A. This is the estimated time from takeoff or departure from an airfield or enroute delay location to the last fix shown in the ROUTE OF FLIGHT, exclusive of planned enroute delays.

Q. How is ETE determined to an alternate airfield?
A. This should include time from the missed approach point to the missed approach holding point, to include one circuit in the holding pattern, and time from the missed approach holding point to the alternate, to include time for approach and landing.

Q. What formula is often used in preflight planning to determine if fuel requirements can be met?
A. TTL TIME x FUEL CONSUMPTION = TTL FUEL - RUNUP = T/O FUEL REQ'D FOR MISSION (lbs);

Total time can be determined by using the "R-E-A-R-A" formula:

FORMULA	EXAMPLE	
(R)unup	0 + 20	(minutes)
(E)nroute (t/o to IAF)	0 + 35	
(A)pproach	0 + 15	
(R)eserve	0 + 30	
(A)lternate	0 + 35	
TOTAL TIME	2 + 15	

TTL TIME x FUEL CONSUMPTION = TTL FUEL - RUNUP = T/O FUEL REQ'D FOR MISSION

(2 + 15) x (550 pph) = 1238 lbs - 184 lbs = 1054 lbs

NOTE: The difference between the computed "t/o fuel req'd for mission" and actual t/o fuel at departure is the fuel surplus. Surplus fuel is important since enroute traffic delays and holding at the destination are not provided for in "R-E-A-R-A" above. Reserve fuel is for UNFORESEEN circumstances. Do not plan to use reserves for routine delays.

Q. What are the general weather requirements for planning an IFR flight?
A. Destination weather must be forecast to be equal to or greater than the published weather planning minimum for the approach procedure to be flown at ETA through 1 hour after ETA. When there are intermittent weather conditions, predominant weather will apply.

Q. Aviators flying helicopters may reduce destination and alternate Category A visibility minimums by ______%, but not less than ______ mile or metric equivalent.
A. 50%; ¼.

Q. Does the above rule apply to approaches labeled "COPTER ONLY"?
A. No, reduction of visibility for approaches labeled "COPTER ONLY" is not authorized.

Q. Can Category II approach procedures be used in destination or alternate weather planning?
A. No.

Q. What weather planning questions should be asked when planning an IFR flight?
A. - "What weather do I need to legally file to my destination?"
- "What weather do I need in order to not need an alternate?"
- "Does my alternate selection, if needed, qualify?"

Q. Under what conditions would an alternate airfield be required when filing IFR to a destination?
A. If any of the following are true:
- Radar is required to execute the approach procedure to be flown.
- Navaids for the instrument approach to be used are unmonitored.
- The predominant weather at the destination is forecast to be less than 400-1 above the weather planning minimum required for the approach to be flown at ETA through 1 hour after ETA.

NOTE: The one exception to the above is that, regardless of the status of radar or navaids, an airfield does not require an alternate if descent from enroute minimum altitude for IFR operation, approach, and landing can be made in VFR conditions.

Q. Careful planning should be involved in the selection of an alternate airfield. What questions should be asked when deciding whether or not an airfield is acceptable for use as an alternate?
A. 1. What is the worst weather condition forecast for the airfield at ETA through 1 hour after ETA? **If you can descend from enroute minimum altitude for IFR operation as well as make the approach and landing in VFR conditions then the airfield may be used as an alternate regardless of the answers to the next four questions.** If the worst weather conditions are less than 400-1 above the weather planning minimum required for the approach to be flown then the airfield may not be used as an alternate.

2. Is the approach procedure to be used at the alternate shown as not authorized (NA) in FLIP? If the answer is YES then the airfield may not be used as an alternate unless the conditions in #1 above apply.

3. Is radar required for the approach procedure to be used at the alternate? If the answer is YES then the airfield may not be used as an alternate unless the conditions in #1 above apply.

4. Are the navaids to be used for the approach procedure at the alternate

---

airfield unmonitored? If the answer is YES, that is, if the navaids are listed in FLIP or NOTAMed as "unmto" then the airfield may not be used as an alternate unless the conditions in #1 above apply.

5. Is a control zone in effect at the alternate? If a control zone does not exist or is not in effect at the airfield then it may not be used as an alternate unless the conditions in #1 above apply.

MEMORY AID: For determining if airfield is suitable as an alternate:

W - weather 400-1 above planning minima
A - no "A(NA)"
R - radar not required
N - navaids monitored
C - control zone in effect

Use the same acronym minus the "A" and "C" for determining if an alternate is required.

W - weather less than 400-1 above planning minima
R - radar required for approach to be flown
N - navaids to be used for approach are unmonitored

LISTED BELOW ARE TWO SAMPLE CHECKLISTS FOR PLANNING AN IFR FLIGHT. THE FIRST ONE IS FROM FM 1-240, APP B. THE SECOND CHECKLIST IS A LITTLE MORE COMPREHENSIVE AND IS A COMPOSITE FROM SEVERAL SOURCES (AUTHORS UNKNOWN).

FROM FM 1-240, APP B:

*Initial Weather Check
- Takeoff
- Destination and Alternate
- Enroute
- Winds aloft and temperature
- Freezing level, thunderstorms, ice, and turbulence

* Preparation
- Check aircraft equipment
- Check enroute supplement for departure, destination, and alternate airfields
- Check enroute chart for control zone effective hours
- Check NOTAMs - Class I and II

*Planning
- Review SID departures and approach plates, if available, for departure airfield
- Review enroute chart. Consider departures and arrival procedure when determining flight route
- Fill in flight plan information, except TAS, time, and distance
- Consult the operator's manual. Compute PPC.
- Compute flight log

IFR FLIGHT PLANNING

- Complete flight plan entries
- Receive weather briefing
- File flight plan
- Brief copilot
- Conduct preflight according to current checklist

IFR PLANNING CHECKLIST

[ ] Check local Reading File
[ ] Check ALSE up-to-date
[ ] Check for type of NAV equipment available (VOR, NDB, DME, GPS, etc)
[ ] Verify that all publications/maps to be used are current
[ ] Check approach plates:
    [ ] Type of approaches and minimum weather required at destination
    [ ] Type equipment required to execute approach
    [ ] Alternate not authorized, departure not standard, etc
    [ ] Weather required at departure location & departure procedures/SIDs
[ ] Check IFR Supplement:
    [ ] Check destination airfield (avail of fuel, effective CZ hours, PPR required, etc)
    [ ] Check departure airfield
    [ ] Check all navaids to be used (range, "unmonitored", dead spots, etc)
[ ] Check NOTAMS:
    [ ] Class I
    [ ] Class II
    [ ] Distant, if civilian (ctc any FSS)
    [ ] Local, if civilian (ctc FSS servicing airfield)
[ ] Get pre-flight weather briefing - general forecast/outlook for:
    [ ] Departure location/Enroute/Destination
    [ ] Winds aloft and temperature (use approx times for departure and arrival at destination)
[ ] Determine if weather requirements can be met for:
    [ ] Departure
    [ ] Destination
        ***** Alternate required?   YES [ ]   NO [ ]
[ ] Plan route of flight
    [ ] Check approach plates for approach to be used and use of transitions at destinations
    [ ] Check approach plates for departure procedures (where does it take you?)
    [ ] Use the enroute chart to plan the route from takeoff procedure to transition fix
[ ] Fill out DD Form 175
[ ] Plan altitude to be flown
[ ] Find True Airspeed (performance charts)
[ ] Fill out flight log:
    [ ] Check points, times, distances, headings

  - [ ] Use winds aloft forecast and the TAS to determine ground speed
  - [ ] Use ground speed to find ETE for each leg on flight log
- [ ] Fill out ETE on flight plan
- [ ] Fill out or verify DD Form 365-4 (Weight & Balance)
- [ ] Prepare PPC
- [ ] Plan fuel required to fly the mission as planned (see formula above)
- [ ] Obtain official weather briefing (DD Form 175-1)
- [ ] Verify Mission Brief and Risk Assessment are completed
- [ ] Aircraft keys/cards/required publications
- [ ] Preflight using appropriate checklist
- [ ] Crewbrief - Crew Coordination

REFERENCES

ATTITUDE INSTRUMENT FLYING/BASIC INSTRUMENTS

ATTITUDE INSTRUMENT FLYING, SH 2/5/9/9E/10-621-2.5, JUNE 1991, USAAVNC.

BASIC INSTRUMENTS, ROTARY WING, PT 102, FEBRUARY 1990, USAAVNC.

MAGNETIC COMPASS, SH 2/5/9/9E/10-4262-2, NOVEMBER 1991, USAAVNC.

WIND EFFECT AND VARIATION-DEVIATION, SH 2/5/9/9E/10-4285-2, MARCH 1992, USAAVNC.

NAVIGATIONAL COMPUTER SLIDE RULE, SH 2/5/9/9E/10-4287-3, MARCH 1992, USAAVNC.

NAVIGATIONAL COMPUTER WIND FACE, SH 2/5/9/9E/10-4288-2, MAY 1991, USAAVNC.

ENROUTE LOW ALTITUDE (ELA) CHARTS AND STANDARD INSTRUMENT DEPARTURES (SIDs)

ENROUTE LOW ALTITUDE (ELA) CHART AND STANDARD INSTRUMENT DEPARTURES (SIDs), SH 2/5/9/9E/10-9291-3, FEBRUARY 1992, USAAVNC.

DEPARTMENT OF DEFENSE FLIGHT INFORMATION PUBLICATIONS (DOD FLIPs)

DEPARTMENT OF DEFENSE FLIGHT INFORMATION PUBLICATIONS (DOD FLIP), SH 2/5/9/9E/10-9294-5, PE-1, JANUARY 1992, USAAVNC.

DEPARTMENT OF DEFENSE FLIGHT INFORMATION PUBLICATIONS (DOD FLIP), SH 2/5/9/9E/10-9294-5, PE-2, JANUARY 1992, USAAVNC.

DEPARTMENT OF DEFENSE FLIGHT INFORMATION PUBLICATIONS (DOD FLIP), SH 2/5/9/9E/10-9294-5, PE-3, JANUARY 1992, USAAVNC.

DEPARTMENT OF DEFENSE FLIGHT INFORMATION PUBLICATIONS (DOD FLIP), SH 2/5/9/9E/10-9294-5, PE-4, JANUARY 1992, USAAVNC.

FLIGHT PLANS

GENERAL PLANNING (GP), CHAPTER 4, 10 NOV 94, DEFENSE MAPPING AGENCY.

INTRODUCTION TO RADIO NAVIGATION

INTRODUCTION TO RADIO NAVIGATION, SH 2/5/9/9E/10-9284-1, MARCH 1991, USAAVNC.

RADIO MAGNETIC INDICATOR (RMI)

RADIO MAGNETIC INDICATOR (RMI), SH 2/5/9/9E/10-9286-7, MARCH 1991, USAAVNC.

RADIO MAGNETIC INDICATOR (RMI), PE 2/5/9/9E/10-9286-7, MARCH 1991, USAAVNC.

COURSE DEVIATION INDICATOR (CDI) AND VOR PROCEDURES (enroute)

RADIO NAVIGATION (ID 1347), SH 2/5/9/9E/10-9289-3, MARCH 1992, USAAVNC.

RADIO NAVIGATION (ID 1347), PE 2/5/9/9E/10-9289-3, MARCH 1992, USAAVNC.

VOR PROCEDURES (EN ROUTE), USAAVNC PT 1951, JANUARY 1991, USAAVNC.

HOLDING

HOLDING PROCEDURES, SH 2/5/9/9E/10-608-3, SEPTEMBER 1990, USAAVNC.

HOLDING, USAAVNC PT 1954A, JANUARY 1991, USAAVNC.

INSTRUMENT APPROACH PROCEDURES (IAPs)

INSTRUMENT APPROACH PROCEDURES, SH 2/5/9/9E/10-612-9, APRIL 1992, USAAVNC.

INSTRUMENT APPROACH PROCEDURES, SH 2/5/9/9E/10-612-9, PE-1, APRIL 1992, USAAVNC.

INSTRUMENT APPROACH PROCEDURES, SH 2/5/9/9E/10-612-9, PE-2, APRIL 1992, USAAVNC.

DOD TERMINAL FLIPS, IAP VOLUMES (misc), 2 FEB 95, DEFENSE MAPPING AGENCY.

GENERAL AND INSTRUMENT FLIGHT RULES

GENERAL AND INSTRUMENT FLIGHT RULES FOR ROTARY WING AIRCRAFT, SH 2/5/9/9E-9225-3, APRIL 1992, USAAVNC.

GENERAL AND INSTRUMENT FLIGHT RULES FOR ROTARY WING AIRCRAFT, PE 2/5/9/9E-9225-3, APRIL 1992, USAAVNC.

GENERAL AND INSTRUMENT FLIGHT RULES FOR ROTARY WING AIRCRAFT, SQ 2/5/9/9E-9225-3, APRIL 1992, USAAVNC.

AR 95-1, FLIGHT REGULATIONS, 30 MAY 90, HQDA.

AR 95-3, AVIATION: GENERAL PROVISIONS, TRAINING, STANDARDIZATION, AND RESOURCE MANAGEMENT, CHPT 6, "WEIGHT AND BALANCE", 27 SEP 90, HQDA.

GENERAL PLANNING, (GP) 10 NOV 94, DEFENSE MAPPING AGENCY.

AREA PLANNING, AP/1/1A, 13 OCT 94, AP/1B, 2 FEB 95, DEFENSE MAPPING AGENCY.

REFERENCES

FLIGHT INFORMATION HANDBOOK, 10 NOV 94, DEFENSE MAPPING AGENCY.

IFR-SUPPLEMENT, 2 FEB 95, DEFENSE MAPPING AGENCY.

AIRSPACE STRUCTURE, ATC FACILITIES, AND IFR COMMUNICATIONS

AREA PLANNING, AP/1, 13 OCT 94, DEFENSE MAPPING AGENCY.

ATC FACILITIES AND CLEARANCES, SH 2/5/9/9E/10-618-1, AUGUST 1991, USAAVNC.

IFR COMMUNICATIONS, SH 2/5/9/9E/10-623-6, APRIL 1992, USAAVNC.

IFR COMMUNICATIONS, PE 2/5/9/9E/10-623-6, APRIL 1992, USAAVNC.

IFR FLIGHT PLANNING

IFR FLIGHT PLANNING, SH 2/5/9/9E-649-15, MAY 1992, USAAVNC.

IFR FLIGHT PLANNING, SH 2/5/9/9E-649-15, PE-1, MAY 1992, USAAVNC.

IFR FLIGHT PLANNING, SH 2/5/9/9E-649-15, PE-2, MAY 1992, USAAVNC.

FM 1-240, INSTRUMENT FLYING AND NAVIGATION FOR ARMY AVIATORS, APP B, "IFR FLIGHT PLANNING", 15 DEC 84, HQDA.

COMMENTS/CORRECTIONS

It is my intention that this study guide be as accurate and up-to-date as possible. I understand however, that mistakes can and will be made during the compilation of such material and welcome any comments and/or corrections of a professional nature. If you would like to recommend changes/corrections, please use the following format:

(1) Name

(2) Address and/or phone

(3) Subject area concerned/page (in study manual)

(4) Question/Answer as it appears in this study guide

(5) Question/Answer as it should appear

(6) Reference source for correction (incl page & paragraph)

(7) Additional comments

Send correspondence in care of the following:

Snail Mail:

Bagwell & Associates
329 N. Dogwood Rd
Powell, TN 37849

Email:

bagwell@bandabooks.com

All correspondence will receive a reply.

Thank You!

Bagwell and Associates
329 North Dogwood Rd
Powell, TN 37849

Please send me ______ copies of:

Flight Training Academic Enhancer - Book I (Primary), 1st Edition, (c) 1992 by D. W. Creech

Please send me ______ copies of:

Flight Training Academic Enhancer - Book II (Instruments), 1st Edition, (c) 1995 by D. W. Creech

Enclosed is a check or money order in the amount of

$16.95 per copy

(Price includes tax, shipping, and handling)

Address to be shipped to:

____

____

____

____

Amount Enclosed: ________

Make checks payable to: Bagwell and Associates

** Please allow 4-6 weeks for delivery **

See us at:
http://www.bandabooks.com